Farriery – Foal to Racehorse

SIMON CURTIS, FWCF

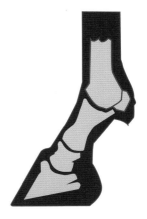

Copyright

© 1999 by Simon Curtis

c/o R&W Publications (Newmarket) Ltd,
5 Kings Court, Willie Snaith Road,
Newmarket, Suffolk CB8 7SG, UK

All rights reserved. No part of this publication may be reproduced, stored in a retrieval system or transmitted in any form or by any means without prior permission from the copyright owner.

First published 1999

ISBN 1 899772 04 9

Published by Newmarket Farrier Consultancy

Typeset and Design by
R & W Publications (Newmarket) Ltd

Printed in Great Britain by Quality Print Services (Anglia) Ltd, Ipswich, Suffolk

To my darling wife, Beverley

Author

The author was born in 1956 in the racing town of Newmarket. His family have been farriers for at least 4 generations. In the middle of the 19th century his great grandfather was a farrier and blacksmith in a village just outside Newmarket. The author's maternal great grandfather came to the town as a groom and later became racing correspondent for the Times of London.

(Left to right) Oliver (Jack) Curtis (Grandfather), Oliver Curtis (Great Grandfather), Will Curtis (Great Uncle) outside the family forge near Newmarket (circa 1905).

In 1972 Simon Curtis began his apprenticeship in the family business of O A Curtis and Sons, where he gained most of his experience in Thoroughbred racing yards and stud farms. The Curtis forge is situated adjacent to Newmarket Heath on the Moulton road. Since his father bought their present forge in 1949 the Curtis's and their staff have shod many of the best racehorses including 11 Epsom Derby winners. Among the many racehorses that the author has been associated with is the great 'Pebbles', Winner of 5 Group 1 races including the 1,000 Guineas and the first English trained Breeders' Cup winner. In recent years the author has concentrated on the remedial shoeing of horses in training and corrective farriery at the stud farm. He has trained 15 apprentices, many of whom are successfully continuing within racing, and one has already plated a Derby winner. In 1992 he founded a farriery clinic attached to the Beaufort Cottage Equine Hospital to which all types of horses with hoof and limb problems are referred.

In 1983 he passed the Diploma of the Worshipful Company of Farriers (Dip.WCF). In 1985 he won a WCF National Educational Award for his paper 'The Foal to Three Year Old, Problems Experienced by the Farrier'. He used the prize to visit and study at The Animal Hospital in Helsingborg, Sweden. In 1987 he became an Associate of the Worshipful Company of Farriers (AWCF) and, in 1990, a Fellow of the Worshipful Company of Farriers by examination (FWCF), with a thesis which

(Left to right) Russel, Peter (uncles), Oliver (Jack) (Grandfather), Don (father), Maurice (uncle) outside the forge, Moulton Road, Newmarket (circa 1955).

was entitled 'The Use of Medial Extensions in the Correction of Carpal Valgus Deformity in Foals'. He has been published in Australia, the UK and the USA, in journals including; Equine Veterinary Education; The American Farriers Journal, Forge, Equine Athlete, Veterinary Practice and Hoofcare and Lameness.

In 1988 the author organised the first International Farriery and Lameness Seminar in Newmarket, England. The seminar was repeated in 1990, 1992 and 1994 attracting, at the last event, farriers and veterinarians from 17 countries around the world. Simon Curtis is a much sought after lecturer and clinician who has taught and demonstrated farriery in many countries including Australia, Canada, Cyprus, Dubai, England, France, India, Northern Ireland, The Republic of Ireland, Scotland, Sweden, and the USA. He has been a lecturer at the Animal Health Trust and Thoroughbred Breeders Association Stud Management Course annually since 1988; The National Stud Course since 1991; and the Worshipful Company of Farriers 'Foot Balance Course' since 1992. He has been an examiner for the Worshipful Company of Farriers since 1991.

Nick, Simon and Mark Curtis outside the forge, Moulton Road, Newmarket (1999).

Acknowledgements

When I was asked how long it had taken me to write this book I paraphrased the painter Whistler's answer and said "my whole lifetime"! This may betray my facetious nature but I think it is close to the truth. Everyone is influenced from the very start of their career. Often those influences remain with you for ever. Writing this book would not have been possible without a great deal of input from many people. I would therefore like to acknowledge the following who may not have helped directly with this work but who have had a deep enough influence upon me to have affected my writing: My father Don Curtis, who was so far ahead of his time and who, until the day he died, believed there was always a better way to shoe a horse; Jacques Hollobone who taught me shoemaking in my apprenticeship; Colin Smith who saw farriery as an intellectual as well as skillful pursuit; Ron Ware and Chris Colles for raising my aspirations; Fran Jurga who introduced me to endless opportunities to meet farriers and veterinary surgeons in the United States of America; Doug Butler; whose book I can only hope to supplement; Dennis Oliver for his guidance and encouragement; The Worshipful Company of Farriers, members of both the Examination Board and Foot Balance team who continue to stimulate thoughts regarding all aspects of farriery; and Mark Curtis, my brother, who has always supported me.

All the photographs, in this book, are my own apart from some slides of hoof cracks from Keith Swan of Australia and pictures of tools and my shoemaking by Geoff Evans. Much of the first draft of the text was typed from my longhand by Geraldine Wheatley and read by Tony Demming. Jan Wade and Louise Holder of R&W Publications gave advice on the design, layout and technicalities of publishing.

I would like to thank the following farriers and veterinary surgeons for checking the text and suggesting changes: Scott Kemble, Steve Norman and Sarah Stoneham and, especially, Hans Albrecht, Russel Brownrigg and David Ellis.

Introduction

Thoroughbred racehorses have been bred and raced for over 200 years. Foot care and farriery has long been recognised as playing a major part in their successful training and breeding. A text for farriers, veterinary surgeons and serious students, covering the fundamental principles of good farriery, is long overdue.

This book covers farriery needs and requirements for the Thoroughbred racehorse from its birth, through development as a foal and yearling; preparation of the feet and shoeing for sale to training and racing. Once the racehorse has finished its racing career, farriery plays an important part in prolonging a successful stud life. Conditions such as hoof cracks and laminitis can affect a mare or stallion's soundness and ability to reproduce.

I always envisaged 'Farriery - Foal to Racehorse' as focusing on the Thoroughbred racehorse, beginning with the young foal at stud and continuing through its career as a yearling, to training and finally back on the stud farm as a stallion or broodmare. With few exceptions I have stuck to this chronological order. However, the book begins with a chapter on the principles of hoof balance. 'Balance' is a term used to describe the way the hoof is trimmed and a shoe fitted to the prepared foot. It has become synonymous with ambiguous treatments and sometimes an excuse not to analyse exactly the requirements of good farriery. Hoof balance is described clearly and lucid directions as to its use are given. It was important to lay a foundation for the book. Unless the reader has some understanding of how the author assesses feet and legs they cannot follow either ideas or directions.

Problems of interference encountered during training, eg brushing, scalping or over-reaching, are given consideration, as are shoeing and plating. Other problems, such as abscesses and quarter cracks, are explained in more depth. Seemingly mundane conditions such as thrush or seedy toe can either be dealt with efficiently or allowed to develop into major problems. The principles for shoeing all types of racehorse are essentially the same. However, the style of shoeing and pattern of shoes and plates differ, and these differences are explained.

Chapters are included on the assessment of limb and foot conformation and how farriery skills can be used to help produce the optimum soundness for any individual. There is also advice on how to enhance the yearling prior to sales. No apology is made for a certain amount of repetition in the book. It is impossible to place all the skills and knowledge that a farrier needs into separate neat chapters.

Terminology varies between, and even within, countries. Therefore an attempt has been made to list and clarify many of the words and phrases used by farriers working in the United Kingdom and the United States of America.

We constantly develop new and re-discover old methods that work. Many mares have had breeding careers lengthened by judicious shoeing. It is clearer which foals or yearlings can be helped by corrective shoeing and/or surgery. There is an awareness of what can be achieved by working on all factors involved in producing sound racehorses and of the teamwork needed to succeed. This book aims to cover the options involved in the successful shoeing and trimming of racehorses.

Contents

1.	The principles of hoof balance	1
2.	Caring for the foot of the foal	13
3.	Medio-lateral deformities in foals and yearlings	21
4.	Two methods of creating medio-lateral extensions for foals and yearlings	37
5.	Flexural deformities in the young horse	43
6.	Making a composite and aluminium toe extension shoe	57
7.	A method of treating severe digital hyperextension (flaccid tendons) in foals	61
8.	Preparing yearlings for the sales	67
9.	Shoeing the racehorse in training	77
10.	The racehorse in training: problems associated with imbalance	91
11.	The treatment of hoof cracks and hoof wall lesions	105
12.	Two methods of patching quarter cracks	123
13.	Shoeing mares and stallions at stud	131
14.	The farrier's relationship with the horse, its trainer, owner and veterinary surgeon	141
15.	Common conditions of the foot	147
16.	Forging bar shoes	155
17.	Terminology and technical language	167
	References and further reading	191
	Index	192

CHAPTER 1

FARRIERY – FOAL TO RACEHORSE

The principles of hoof balance

One dictionary definition of balance is 'the harmony of design and proportion'. Good horse-shoeing involves matching, as closely as possible, the foot with the leg so that the shape and proportions of the foot are the most suitable for that limb. The way the horseshoe is attached to the foot can either enhance or denigrate it.

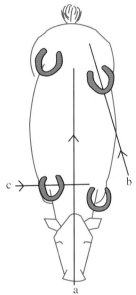

Figure 1.1: Views of static balance: a) looking down the spinal axis of the horse; b) from the front of each individual limb, ie the way the foot is pointing; and c) lateral view.

There are some 'golden rules' in the preparation of a foot and the fitting of a horse-shoe. These rules are ideals and the closer we are to obtaining these ideals, the more likely the horse is to perform to its maximum potential and to remain sound for a prolonged period. There is little doubt in the mind of most farriers and equine veterinary surgeons that a large proportion of the lamenesses that are seen today can be avoided or treated by simple compliance with the rules governing hoof balance. It must, however, be borne in mind that many hoof balance problems are caused by poor leg conformation. The hoof distorts as a result of uneven loading of pressure from above. Judicious shoeing may maintain soundness in horses with poor conformation, but it will not alter the limb of a mature horse. Farriers need to be aware at each shoeing of any hoof distortion so that he can reshape the foot and bring it back into alignment.

Poor conformation (broken-back hoof-pastern axis [HPA]) is usually induced by shoes being left on too long and/or short shoeing. Most owners judge their farrier by whether his shoes stay on and many farriers are pressurised to shoe short. Short-fitted shoes and shoes left on too long eventually cause lameness, which may be permanent. Any hoof imbalance is aggravated by leaving the time between shoeing and/or trimming too long. The hoof grows increasingly out of balance as each day passes.

Many conformational problems are insurmountable, a misaligned limb slowly causes the foot to distort, irrevocably damaging ligaments, joints, the pedal and navicular bones. The horse goes lame in a matter of weeks but the damage has been accumulating for years. Corrective shoeing for these cases is a long-term prospect and often disappointing.

For the horse owner, the best guard against poor conformation and foot imbalance is to start with a horse with good conformation and to employ a farrier who regularly observes the horse's conformation and movement. Farriery is then aimed at maintaining good hoof capsule shape and alignment.

CHAPTER 1

Figure 1.2: The centre line of weightbearing bisects the hoof; the hoof wall is symmetrical.

they show that they hoof is under severe uneven stress or that the hoof is not being adequately nourished in certain areas, eg laminitis creates a distorted hoof wall with growth rings diverging at the heels and often compressed at the proximal dorsal wall.

Anterior view – this means looking at the foot and limb at rest, from the front. The front limbs should be assessed looking down the spinal axis (Fig 1.1a) of the horse (as in most text books) and from the front of each individual limb (ie the way the foot is pointing, Fig 1.1b). A vertical axis through the centre of the cannon bone should bisect the hoof into 2 equal halves. The hoof wall should be at the same angle on both sides (Fig 1.2). The wall should not flare out or run-under (Fig 1.3).

Assessment of balance

The assessment of the balanced foot can be broken down into the lateral and anterior views and into the static and dynamic balance, in other words when the horse is standing still or moving.

Static hoof balance

Static hoof balance is the assessment of the horse whilst it is not moving. The horse must be stood up square so that it is bearing weight evenly on all 4 feet. The horse needs to be stood on a clean, level surface, large enough to allow room around the horse for safe viewing. The horse is assessed as a whole, each leg individually and each leg picked up. The hooves should be appraised for shape and distortion, whether they are over-grown, have growth rings or lesions.

A general assessment of the hoof shape and its proportion to a particular horse must be made. Where there are growth rings and they are parallel, they may just signify a change of environment, diet, or illness. If they diverge or are compressed in one part of the hoof wall

Figure 1.3: An angular limb deformity (ALD) causes the hoof wall to flare one side (right) and run-under the other side (left).

Figure 1.2 depicts an ideal foot although this is not often attainable. If the coronary band is not horizontal it should not be the farriers primary objective to level it. Distortion of the coronary band and bulbs of the heel occur due to movement and uneven compression, just levelling them will usually increase the uneven stress and worsen the distortion.

THE PRINCIPLES OF HOOF BALANCE

Looking at the whole horse from the front tells us about his action and the resulting distortion to the hoof capsule and wear on the shoes. If he is wide in the chest and has offset knees (Fig 1.4) it is likely that he toes in, is flared medially and upright on his lateral wall. He probably paddles when he walks, lands hard on his lateral wall and breaks over on his outside toe. His shoe wear reflects this action, it is worn from the outside branch to the outside toe. If the horse is base-wide and has knock knees (carpal valgus), he will have a hoof capsule that is distorted laterally so that it flares on the lateral wall, is upright or under-run medially and may have the medial bulb shunted proximally (Fig 1.3). His action will dish and he will probably land on the lateral hoof wall, smack down hard on the medial side during the weightbearing phase and break-over on the inside toe. If his uneven landing is severe he will be prone to corns and even quarter cracks.

Figure 1.4: Offset knees cause toe-in (varus deformity).

The 2 examples above follow the rule that 'function follows form' in that the horses conformation makes it move in the way that it does. The secondary effect is that the hoof capsule distorts due to uneven loading, a case of form following function. It does this for several reasons: 1) it is continually growing and can therefore be influenced by pressure away from its correct alignment; 2) it is not firmly fixed to the skeleton, eg hoof wall is attached by way of the laminae which by their nature must be flexible, the coronary band is only attached to the skin; 3) the hoof may wear unevenly; and 4) horn compresses.

The long axis – looking down the long axis of the cannon, pastern and hoof capsule (eyelining) while the leg is held by the cannon as close to the knee as possible and allowed to hang loosely gives us the best guide to medio-lateral balance (Figs 1.5 and 1.6). A 'T' square can be aligned along the back of the cannon to show whether the solar surface is at 90° to the long axis (Fig 1.7).

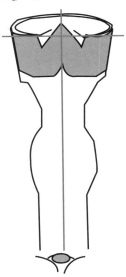

Figure 1.7: Eyelining; looking down the long axis of the cannon, pastern and hoof capsule.

Solar view – looking at the underside of the hoof can tell us a lot about foot balance (Fig 1.8). The frog is the best guide to the foot's symmetry. A trimmed frog is a wedge-shape that starts between the heels and ends in a blunt point (the apex) just forward of the hoof centre. The frog should exactly bisect the foot,

CHAPTER 1

Figure 1.5:
Eyelining; holding the cannon from the inside allows the leg to hang free.

Figure 1.6:
Eyelining along the long axis of the cannon shows medio-lateral deformity and hoof capsule distortion. The medial heel (on the right) is shunted. The lateral hoof wall (on the left) is higher and flared.

THE PRINCIPLES OF HOOF BALANCE

Figure 1.8: The shape of the solar surface shows distortions.

the hoof should be equal in shape and proportion either side of the frog. The shoe should be applied in the same manner.

If the hoof wall and/or the shoe is not evenly proportioned (Fig 1.8) then the farrier should attempt to re-shape the foot and set the shoe symmetrically around the frog, so that it too is bisected equally by the centre of the limb (Fig 1.9).

Lateral view – from the side view (Fig 1.1c) it is essential that the hoof-pastern axis (HPA) is in perfect alignment. Ideally, the hoof wall and angle at the heel should also align (Fig 1.10). If the HPA is broken-back (Fig 1.11), then great strain is thrown upon both the suspensory apparatus and dorsal wall laminae. The suspensory apparatus is the system of tendons and ligaments that allows the horse to lock its limbs and rest standing up. It also protects the horse's fetlock from hyperextension during fast work and jumping.

Failure to maintain the HPA can lead to many lameness conditions including mechanical laminitis (the dorsal wall laminae are torn from the hoof wall), navicular syndrome, caudal hoof lameness, degenerative joint disease (DJD) of the coffin and pastern joints, flexor tendon injuries and dorsal hoof wall lesions.

A broken-forward HPA (Fig 1.12) shows the dorsal hoof wall to be more upright than the pastern axis. Although not as serious as a broken-back HPA this can still lead to stumbling and excessive landing on the heels. A broken-forward HPA with a steep, sometimes concave, dorsal hoof wall is called a club foot (Fig 1.13). The solar shape is similar to a hind foot with wide heels and a well developed frog. An upright foot should not be confused with a club foot. The HPA is in alignment and the solar shape is usually oval and contracted at the heels. The cause of an upright foot, especially if unilateral, may be an injury to that limb at some time.

Dynamic hoof balance

However much we admire our horse standing, it is his movement and soundness that are the greatest tests of balance. If a horse appears correctly balanced when standing but is either unsound or does not move well when ridden, then we need to observe his dynamic balance.

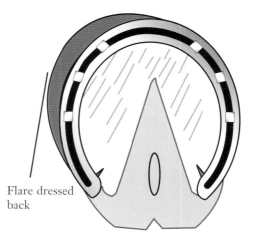

Flare dressed back

Figure 1.9: The shoe is symmetrical to the frog.

CHAPTER 1

Figure 1.15:
Severe anteriorposterior imbalance with broken-back HPA, flared toe and under-run heel. The shoe has also sunk in at the heel.

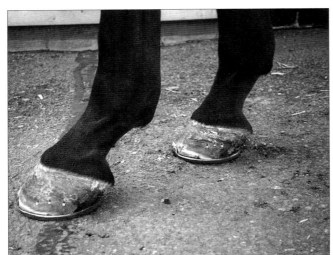

Figure 1.16:
The toe and heels have been dressed back and a shoe set to give adequate caudal support.

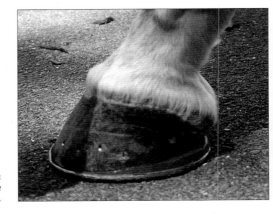

Figure 1.17:
Shoeing with a heel wedge to elevate the heels and a bar shoe for caudal support.

THE PRINCIPLES OF HOOF BALANCE

Farriery cannot influence the limb in the air, all the farrier can do is influence the foot landing, during the weightbearing phase, and taking off (break-over). Almost all hoof balance-related injuries are due to the landing and weightbearing phases of the stride, these can be affected by both foot trimming and shoeing.

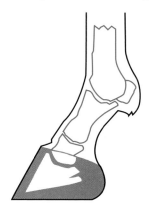

Figure 1.10: Normal (straight) hoof-pastern angle (HPA). The phlanges and the dorsal hoof wall are in alignment, the heel is also ideally in alignment.

Anterior view – stand directly in front of the horse while it is walked towards you. If you squat down low you can watch the instant that the foot strikes the ground. Good medio-lateral (side-to-side) balance is seen when the hoof lands level. If the hoof is out of medio-lateral balance, usually due to poor limb conformation, then one side lands first (contact) and the other side is immediately slammed into the ground (impact) in the weightbearing phase.

A typical example of this is the horse that 'toes out' (Fig 1.3). As his foot lands, the outside toe (which is usually flared) impacts first, as the body weight passes over the limb and foot the inside heel is snapped down. The weight of the horse is rolled over the inside toe as the foot breaks over and lifts off. This type of action is liable to cause the medial (inside) heel to shunt up higher than the lateral (outside) heel. It may also cause corns on the inside and quarter cracks to the hoof wall. The medial heel, because it is shunted up, appears higher at first glance and is often trimmed down to 'level it' with the lateral heel. This leads to greater dynamic imbalance and eventual lameness.

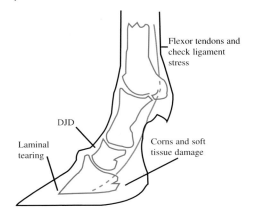

Figure 1.11: Broken-back HPA; failure to maintain the HPA can lead to many lameness conditions.

Lateral view – the observation of gait from a lateral viewpoint can also help in foot trimming. In the well balanced foot the foot should land level or slightly heel first. Horses that land toe first are usually showing signs of pain in the caudal (rear) area of the foot and are prone to stumbling. Where the toe is long and the heels are under-run the gait becomes more animated for the horse to break over. Where there is or has been a laminitic condition the toe will usually flip up in an exaggerated manner and the foot will clearly land heel first.

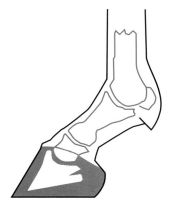

Figure 1.12: broken-forward HPA.

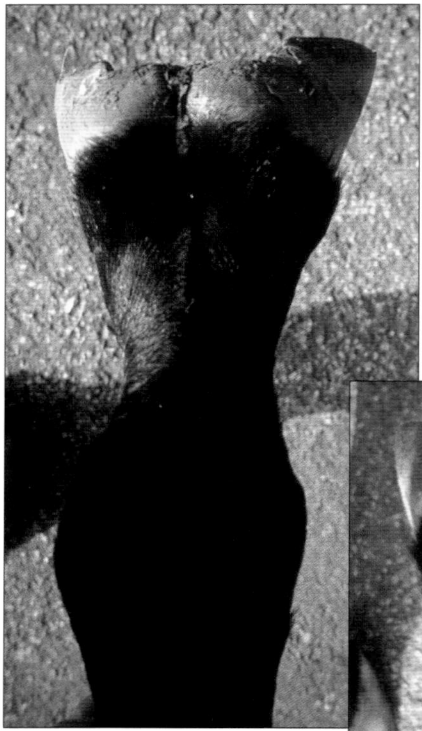

Figure 1.19:
A foot showing high lateral distortion and medial bulb shunting.

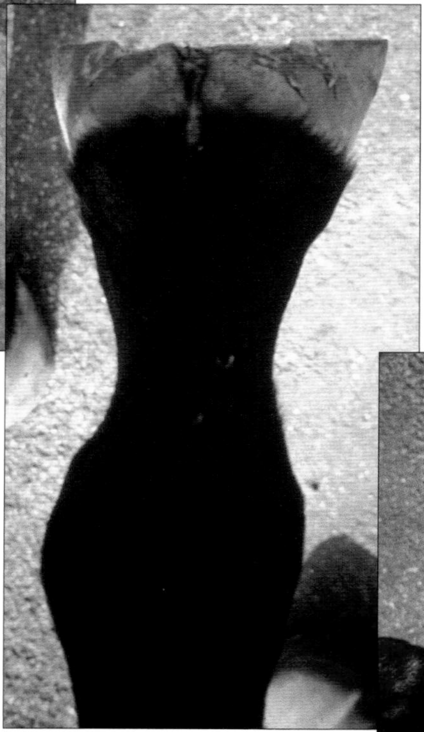

Figure 1.20:
Figure 1.19 after foot dressing to bring back into balance.
The lateral side (on the right) has been lowered so that the distal border of the f[oot] is at 90° to the long axis of the cannon.

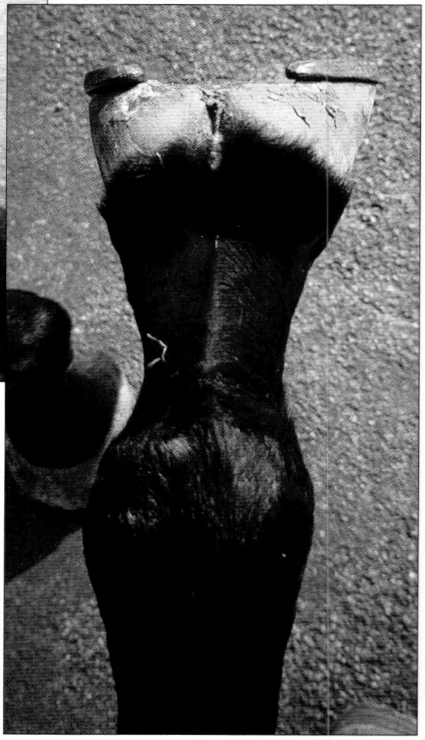

Figure 1.21:
Figure 1.20 after shoeing for medial support. The shoe has been set symmetrically on the foot so that the lateral branch is tight and the medial branch is full (wider).

THE PRINCIPLES OF HOOF BALANCE

Shoeing

Shoeing offers us an additional way of adjusting the balance of the foot. As a general rule the foot will move, in relation to the limb, in the direction that a shoe is extended. To achieve this the foot must be strictly trimmed in accordance with the following guidelines: 1) the solar surface must be at 90° to the long axis of the cannon; 2) the dorsal wall must be dressed back straight to align with the proximal third phalanx; and 3) the heels must be dressed back to their caudal extremity.

Shoeing for anterioposterior balance

If the horse has a broken-back HPA, with a long flared toe and an under-run crushed heel (Fig 1.11), we can dress back the toe and shoe with upright heels and a rolled or rocker toe (Fig 1.14). This will support the suspensory apparatus, the caudal hoof capsule and improve break-over, thereby reducing stress in the dorsal wall (Figs 1.15 and 1.16).

The eggbar shoe has long been used where there is a need to improve a broken-back HPA. It has the additional beneficial effect of giving more rigidity to the hoof capsule by reducing shearing movement. A straight bar will improve rigidity but will not give the same influence in repositioning the foot in relation to the limb and body weight of the horse. The eggbar can be seen as a caudal extension shoe. It is very effective where the hoof trimming is correct and it is applied on the correct principles.

Elevating heels is possible where the HPA cannot be corrected by trimming and shoeing alone. There are various ways to achieve wedging. Shoes can be forged to increase the height at the heels. This takes skill and time and adds to the weight of the shoe. Weld on wedges have been available in recent years. These have the advantage of minimal weight addition and ease of fitting, provided you have a welder. Plastic and aluminium shim wedges are widely used as the easiest and lightest way of adding height. These slip between the shoe and the foot and are usually riveted into position. They have a tendency to wear because of the expansion and contraction of the foot at the heels (Fig 1.17).

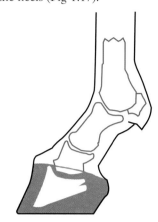

Figure 1.13: A club foot with a steep concave dorsal hoof wall and broken-forward HPA.

Club foot

The club foot (Fig 1.13) is a chronic condition which cannot be cured. It is usually the result of an unresolved flexural deformity involving the deep digital flexor tendon (DDFT). Over trimming of the heels of the club foot may throw such excessive strain upon the flexor tendons and the sub-carpal check ligament that lameness results. That is not to say that the farrier should not attempt to improve the shape. Some judicious trimming of the heels and promotion of toe is advised. Where the dorsal hoof wall is concave it should be dressed back to straight.

One balancing guideline is to trim the foot so that the sum of its dimensions match the sum of the dimensions of the normal foot on the other side, eg if the normal foot is 12.5 x 12.5 cm (5 x 5") and the club foot has a width of 11.5 cm (4 1/2"), trim the heels until the length is 13.5 cm (5 1/2"). The farrier should try to create a straight hoof-pastern axis (HPA). There should not be an attempt to match the feet in shape, it is difficult enough to securely nail to a club foot without trying to shoe cosmetically.

Because of the internal stress to the laminal attachment the club foot is prone to poor quality hoof wall which can cause difficulties when nailing. Seedy toe is often present, this compounds shoeing difficulties.

Upright foot

An upright foot needs alternative shoeing considerations to the club foot. The opposing limb to the upright foot is usually wider and flatter. This may be due to extra weight taken on it when the upright foot was formed by a non-weightbearing injury. The flat foot may continue to carry more weight than the upright foot. If there is a concerted attempt to bring the height of the upright foot down additional strain may be thrown on the flat foot. It is usual to support the caudal area of the flat foot by shoeing with length or possibly an eggbar shoe.

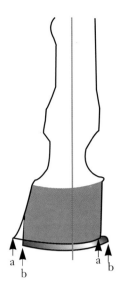

Figure 1.18: A lateral extension shoe helps to equalise the weightbearing through the centre of the leg; arrows a) mark the original extremities of M-L support, arrows b) show the rebalanced extremities.

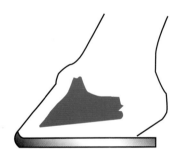

Figure 1.14: The foot in Figure 1.11 has been dressed back to the correct proportions and shod with upright heels and a rocker toe.

Shoeing for medio-lateral balance

In the case of a medio-lateral imbalance where trimming alone will not bring the centre of balance back through the middle of the hoof, the shoe can be set wider to support the limb more evenly (Fig 1.9). A shoe that extends horizontally beyond the distal boarder of the hoof wall is known as a lateral extension shoe (Fig 1.18). This is the generic name, to be precise the name should refer to whether the extension is on the medial or lateral side of the foot. An extension can be anything from just fitting a shoe a few millimetres wide to a 'jug handle' shoe with an extension of more than 10 cm.

Types of lateral extension shoes

There are a number of ways to make a lateral extension shoe. The simplest is just to fit the shoe wide of the foot and box off the edge of the shoe (Fig 1.21). Re-stamping and pritcheling nail-holes further in the webb of the shoe allows for wider fitting with secure nails. Fitting the shoe normally to the foot and then welding an extension to the shoe as required is a simple and accurate method. Using round stock for this method creates a good finish that is not easily pulled off. Hand forging lateral extension shoes tests even the best shoemakers. Either the extension branch is widened by upsetting or the normal branch is drawn down.

Fitting lateral extension shoes

A good rule of thumb when fitting an extension is that once it comes out from the foot it stays out, ie if it begins to come out from the toe quarter and extends 5 mm, then it stays out at least that distance until it passes the heel. There are many formulae given for calculating the width of an extension. Most of them are based on theoretical geometric shapes which look very good on paper but are impractical in real shoeing. If a mature ridden horse requires a medial extension we are constrained by the fact that it may strike the opposing limb and that it may easily be pulled off. In the majority of cases the extension should equalise the distance from the outer edge of each branch to the centre of the foot (a line anterioposterior through the frog) (Fig 1.9). In severe cases where a lateral extension is required then it is possible to extend out so that the centre of weightbearing equally bisects the medial and lateral margins of the shoe (Figs 1.3, 1.19, 1.20 and 1.21).

CHAPTER 2

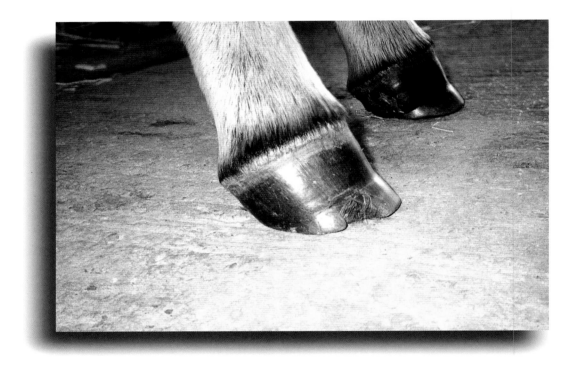

Caring for the foot of the foal

How often we hear that today there are more leg and foot problems than ever before. Is it that we are more aware of conformation at an early age and spot these deviations, or are there some underlying factors that have brought about a perceptible deterioration? The truth is to be found somewhere in between. Most of us are becoming more aware and more critical of any faults, but there are several reasons why we are seeing more deviations.

Breeding

By and large, racing avoids the breeding fashions associated with show types. Once an animal is put in a show ring and judged on a whim the breed is on the way downhill. That is not a worry in racing you might say. We breed from horses proven by the exacting demands of the racecourse. That may be true of one half of the bloodline but it is not necessarily true of the female side (Fig 2.1). To improve, or even just maintain the breed, racing should prove soundness as well as speed and stamina.

The male side is usually removed from the bloodline if unable to produce good results. To have stayed sound long enough to have achieved the level of fitness required to compete at the top must demonstrate a certain quality of soundness. However, the female may have such poor conformation that she never even gets into training let alone stand its rigours. A filly that breaks down in training and never races is surely less successful than one finishing sound but last in a seller. The crux of the matter is that there is little culling of an individual however poor her conformation might be. The generation upon generation effect of this must surely show an increase in the number of deviations in our foals and yearlings.

Nutrition

Feeding has been closely linked to developmental orthopaedic disease (DOD) and, consequently, angular limb deformity (ALD). Flexor tendon deformities (FDs) of the contracted tendon/club foot type may also be brought about by uneven growth patterns. Although a great deal more is known about nutrition these days, the effects are still a long way from being understood completely.

Environment

Any rigid diet is thrown by a sudden change in climate and/or environment. A stud farm may be able to control feeding but it is unable to control the rainfall and hours of sunshine. These affect the grass and consequently food intake. Whether the ground is hard or soft or abrasive may have an influence upon limb development.

Farriery

The growth in the numbers and value of racehorses world-wide has led to a realisation of the importance of good farriery. We now need to get away from the bland platitudes such as 'no hoof no horse' and move on to a more scientific understanding of the inter-relationships of the foot and the leg. Farriery has passed through something of a renaissance during the 70s and 80s. The farrier's ability to correct deviations has often been clouded in myth. Are the problems of ALD and FD caused by negligent or poor farriery? Can foot dressing or shoeing bring about radical changes to conformation?

It is reasonable to expect that a good farrier is capable of keeping a young and well conformed

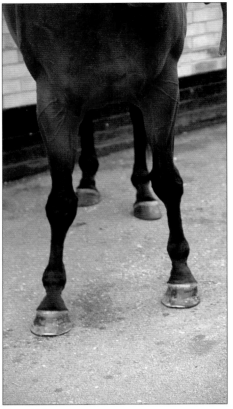

Figure 2.1:
A well bred unraced mare showing severe offset knees and fetlock varus.

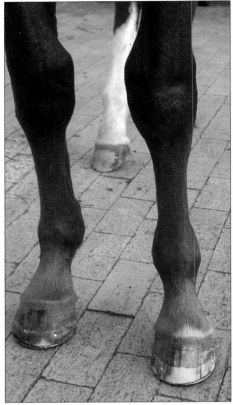

Figure 2.2:
2-year-old with carpal valgus and fetlock varus. It later suffered from a knee chip and a quarter crack.

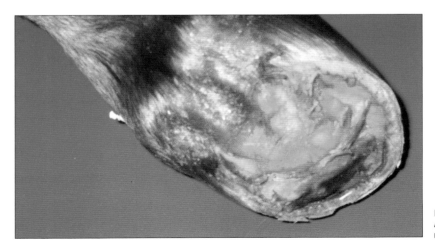

Figure 2.3:
A newly born foal's unhardened foot.

horse straight. With some understanding of physiology and especially growth plates, minor deviations can be corrected. What of the more serious problems? The foal whose heels are 2 cm off the ground, the yearling with a deviation at the knee or the foal who has a rotated limb? We need to know if these are treatable and we need to know if in attempting to correct them we might cause another problem. The answer to the yearling that deviates outwards from the knee is not to rotate the hoof capsule inwards.

Veterinary

Veterinary surgeons are often involved in a stud farm's attempt to avoid and treat limb problems. Their role is one of guidance unless surgical procedures are called for. If the vet insists on trimming a foal or yearling himself (it does happen) then you have to consider that either they believe they can trim better than a farrier or that they believe that the farrier is so inept that they are trying to cancel out or correct the farrier's work. Whether the veterinary surgeon is right or wrong, there is a problem.

In recent years this type of conflict between veterinary surgeon and farrier has lessened to be replaced by mutual respect for each other's skills and abilities. The author has always found that the more a farrier communicates with the veterinary surgeon then the more support they will give you.

Farriery intervention

The argument "Whoever saw the perfect specimen?" and "There are plenty of horses with poor conformation that win races" is usually an excuse for apathy. Some horses win despite our erroneous attempts through ignorance or neglect to slow them down. That does not mean that we should not continually attempt to produce good conformation. Maybe that poorly conformed but good handicapper could have been a stakes winner? Many good horses have had their careers shortened by injury related to conformation or shoeing that could have been avoided as foals (Fig 2.2).

Ensuring that the foal develops healthy feet in balance with its limbs is an essential part of the production of a sound racehorse. Whether the foal is bred to sell or race should be incidental to this aim. From Simon of Athens 3,000 years ago to present day, a library of books have been written upon the subject of farriery and its effects on limb disease and conformation. Unfortunately, very little has been written about youngstock with regard to limb deformity and corrective farriery. There are many misconceptions about what a skilled farrier is or is not able to achieve. All too often a farrier is stopped from improving foot and limb conformations by stud farms and their advisers because they do not believe that the farrier can achieve anything. At the same time he is asked to carry out work upon another foal that he cannot help and which may well create other problems. Part of a farrier's skill must be to demonstrate and persuade by the success of his methods and the failure of others; it takes time, usually years.

To get the best from their craft, a farrier needs to have some understanding of what they can achieve and also their limitations. They should also be aware of the causes of many of the conditions that affect foals. The standard of farriery work on stud farms is most influenced by the farm management's personal interest, the level of environmental conditions, and the quality of handling.

Hoof shape

The foot and especially the hoof are more affected by dynamics than the rest of the limb. We tend to recognise deviations and deformities by their effect upon hoof shape and presentation. This leads us to forget that changes to the foot are a secondary condition. Because of this we often aim our treatment at the symptom rather than the cause. Thus a contracture of the deep digital flexor tendon

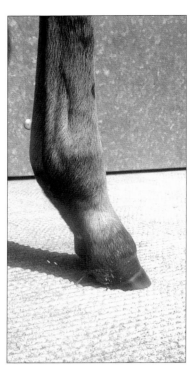

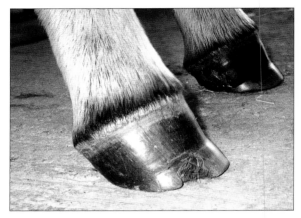

Figure 2.4:
Foal foot. A foal with a flexor tendon deformity showing the different hoof wall angle of the original hoof (distal) and the new hoof since birth (proximal).

Figure 2.5:
The foal foot is often broken off or chipped as it grows nearer to the distal border.

Figure 2.6:
Viewing a foal on a level walking surface.

(DDFT) results in a club foot. We aim all our efforts at trimming the heels down and levering them to the ground. Or we describe a foal as 'toeing-out' and forget that it does not so much toe-out as turn its whole limb out. The foot is merely following the limb. Attempts to turn the toe back which leave the limb still out only gives rise to further complications.

Foals are born with symmetrical and well balanced feet. Any alterations that occur later are invariably due to uneven weightbearing. The foal's foot changes rapidly in response to stress and attempts at reversing this process are often disappointing.

Because of the construction of a horse's foot we can attach orthopaedic devices (corrective shoes) to it that influence the limb. This should not be confused with normal farriery which is basically just reducing the excess growth and rebalancing uneven wear. It should be understood that even a small deviation in the limb will produce uneven wear and growth. If neglected the foot will become distorted and this in turn will create a greater imbalance of load through the limb. Regular hoof trimming reduces the incidence of deformities.

Normal trimming

A foal is usually trimmed at about one month of age unless there is a reason for earlier attention (see following chapters). The rate of hoof growth in a foal is considerably faster than for a mature horse or even a yearling. A foal grows about 15 mm per month, a yearling 12 mm per month, and a mature horse 9 mm per month.

The foal is born with an undeveloped foot proportionate to it's size and relating to it's bone thickness (Fig 2.3). Since the coffin bone is rapidly enlarging, the hoof is growing in 2 ways; it is lengthening (15 mm per month) and it is expanding. With rapid growth of both types the foal frequently has a hoof capsule that is wider proximally than distally. Normal trimming is aimed at widening the foot and bringing back the heels. If the heels are not brought back by trimming then they will begin to collapse and under-run.

In most cases sole trimming is inappropriate and unnecessary. The frog usually needs trimming up on both sulci. The bars are only trimmed if prominent. The heels should not be opened. Rounding off is necessary to prevent breaks and splitting.

Foal foot

The foal foot is the term used to describe the projection of the dorsodistal hoof wall that becomes apparent at around 4 months (Fig 2.4). This would appear to be the remnants of the original hoof capsule. This was produced under non-weightbearing conditions, within the uterus. After birth the hoof bears weight and the new horn produced is affected by greater stress. There is then a clear change in the angle of the dorsal wall.

There comes a point where the foal foot must be trimmed off. If left it can break (Fig 2.5). The timing of its removal has to be on an individual basis. Where there is a flexural deformity involving the deep digital flexor tendon (DDFT) the foal foot becomes very pronounced. Trimming off during this condition can allow the heels to elevate.

Techniques used in applying shoes to young horses

When applying a shoe to a young horse several factors must be considered.

Weight: It is a golden rule of all horse shoeing that the minimum weight of shoe should always be used. Young horses in paddocks or box rest are not in abrasive conditions where wear is a factor. Therefore steel shoes are unnecessary. Aluminium should be used instead. All the present types of glue-on shoes are plastic and light (Figs 5.21 and 5.22). Where a rigid extension is required then aluminium should be used. Acrylic filler above the extension will strengthen it and spread stress (Fig 5.23).

Nailing: Nailing shoes on young horses presents many difficulties. They have thinner walls in which to nail. The wall may already be broken and/or excessively worn because of a pathological condition. It may be impossible or at least very difficult to nail without pricking (puncturing the sensitive tissue) or nail binding (the nail pressing on the sensitive tissue causing pain). When shoe loss occurs, further damage may result. Also there is always a risk of injury from standing back down on nails.

Behaviour: Young horses are more fractious than mature horses. Few will stand for nailing. Unless they stand still the farrier cannot concentrate completely on his job. Foals should be tranquillised as routine, eg Domosedan removes risk of injury to the farrier, handlers and the foal.

Restricting hoof movement and growth: The hoof wall grows downwards from the coronary band. In the young horse the whole foot is expanding in size due to growth. The foot of any horse is flexible especially laterally from the quarters caudal. Hoof movement plays an important part in reducing concussion to the limb. When shoes are attached with nails it should not be assumed that these do not restrict the hoof. With normal hoof movement restricted the foot may contract. A normal shoeing guideline for mature horses is that nails are positioned back no further than the widest part (quarters). This is an attempt to allow normal movement. Slow motion film of the foot (C. Pollitt, personal communication) shows that the whole of the foot is distorted during the weightbearing phase of the stride. Nails within the hoof of a young horse must make the wall more rigid and therefore compromise foot growth. The 'cuff' type glue-on shoe (Dalric, Dalmer, Germany) is very restrictive. There are strong manufacturer's recommendations as to the length of time that these should be left on. The 'tab' type glue-on shoes (Baby Glu, Mustad Hoofcare SA, Bulle, Switzerland) do not restrict the foot in any way.

Skills and materials: Skills of the highest level are needed for a farrier to make and fit shoes to foals. The techniques needed to use glue-on shoes require farriers who are committed enough to learn completely different skills. They also must attend courses and conferences and work with an experienced farrier. Finally, there has to be considerable investment in different materials and tools.

The techniques and skills needed to prepare the foot of a foal correctly, to make a shoe and nail or glue-on a shoe are of a very different order from those which most farriers use in the normal course of their work. Pressurising farriers to undertake this type of work, when they do not have the skill or desire to carry it out, will inevitably prove unsatisfactory to all parties.

The veterinary surgeon can help the farrier by protecting him from unreasonable client expectations. Many clients see farriery as an 'easy fix'. They believe that trimming or shoeing alone, without changes in exercise and nutrition, will achieve results. The veterinary surgeon should also be willing to give tranquillisers when shoeing is undertaken.

CHAPTER 3

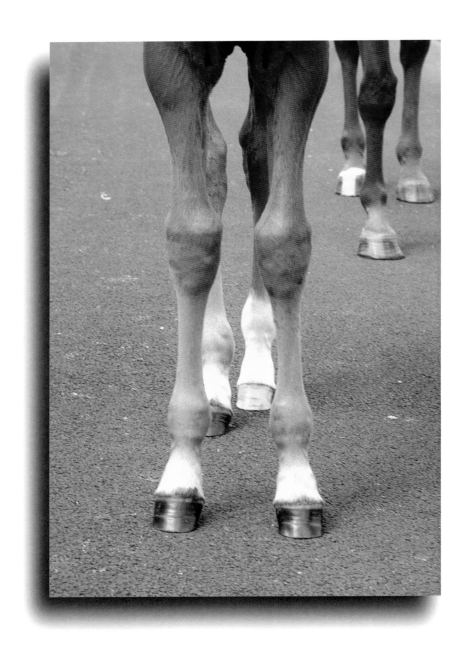

Medio-lateral deformities in foals and yearlings

On many stud farms medio-lateral foot/limb conformation is assessed simply as 'toeing-in' or 'toeing-out'. This leads to an over-simplification of the treatment regime required to bring about an improvement. Radical over-lowering of one side of the hoof capsule distorts it without necessarily bringing about an improvement to the conformation. Early recognition of the type of deformity and its site, together with the age of the foal, will give a clearer diagnosis and a better prognosis.

This chapter describes how to assess and treat medio-lateral deformities in foals on Thoroughbred stud farms. The farriery procedures described are not designed to create a cosmetic change (ie for selling). Rather, they are aimed at reducing uneven stresses to the hoof capsule and the growth plates within the limb by improving alignment and gait. In Chapter 8 'Preparing yearlings for sales', cosmetic correction is covered.

Assessment of medio-lateral deformities

Foals should be regularly (each month) examined at the walk and standing (Figs 3.1 and 3.2). Careful observation of the gait is necessary to ascertain the position and type of any deformity. Eyelining (looking along the long axis of the cannon, the pastern and the hoof) and flexion of joints is useful in diagnosis. Age must be allowed for when deciding prognosis.

A foal should be examined on a level, clear surface where it can be safely viewed and walked. Walking exaggerates any deformity in the weightbearing phase. It is important that the foal is walked straight and that the head is in line with the spine. The hind limbs are viewed as the foal is walked away from the assessor. Then the foal is turned and the front limbs and gait are viewed as it walks towards the assessor. The flight of the limb and placement of the foot in relationship to the body should be given attention. If there is any doubt in the assessor's mind, the foal should be walked again, when a single limb can be examined exclusively.

The foal should then be evaluated in a static position. Careful attention should be given to limb alignment and the hoof capsule distortion. The foal needs to be stood square with the head forward so that weight is borne equally by the limbs. The limbs should not just be observed from directly in front of the foal (ie along the spinal axis). If they are observed from in front of each individual limb a different picture is gained (Figs 3.3 and 3.4).

Finally the foot and limb should be assessed by eyelining. This is most safely done in the foal's loose box. The leg should be lifted and held at the cannon from the inside of the leg adjacent to the knee. As closely as possible the foot is viewed down the long axis of the limb (Fig 1.4). By lifting the toe of the foot additional information is gained about joint alignment. The foot should be allowed to hang loosely to identify the solar plane of the foot in relation to the long axis of the cannon or the pastern. Distortion of the hoof capsule can be assessed at this point (Figs 3.5, 3.6 and 3.7). Particular attention should be paid to the bulbs to determine if the medial side is shunted (shunting is the term used to describe when one heel bulb is higher than the other). Using the frog as a guide to the centre, and alignment

CHAPTER 3

Figure 3.1:
Walking the foal to assess medio-lateral deformity in the hind legs and feet.

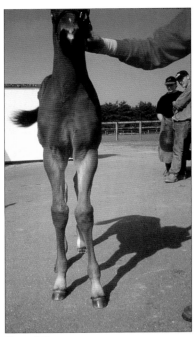

Figure 3.2:
Standing the foal to assess medio-lateral deformity in the front legs and feet.

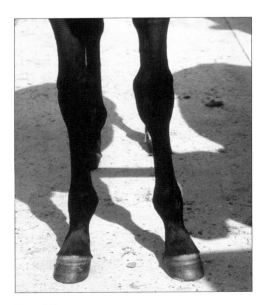

Figure 3.3:
Viewing a foal only from in front – down the spinal axis (Fig 1.1) - can give a misleading assessment. From this view the foal appears to toe-out.

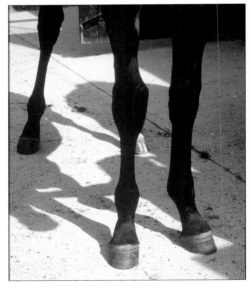

Figure 3.4:
The foal in **Figure 3.3** is now viewed from in front of the leg. The leg and foot can be seen to be in alignment from the carpus down.

of the distal phalanx, the outline of the solar margin of the hoof wall can be assessed for wear or excessive growth, flaring, under-running, and straightening.

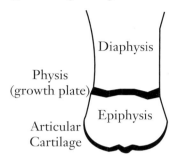

Figure 3.8: A long bone grows in length by a proliferation of cells at the growth plate.

Growth plate	Age at closure	Rapid growth period
Distal tibia	24 months	0-6 months
Distal radius	24-30 months	0-8 months
Distal third metacarpal	12 months	0-3 months
Proximal proximal phalanx	6 months	
Proximal middle phalanx	6 months	

Figure 3.9: Growth plate table; showing closure times and rapid growth.

The knowledge of the foal's age allows a prognosis to be made. Growth plate tables (Fig 3.9; where angular limb deformity [ALD] is being assessed) can be misleading in that they relate to the final fusion. In practice, the ability to influence growth plates stops well before the times given. As a rough guide it is difficult to alter growth plate alignment after 3 months at the proximal phalanx, 6 months at the distal large metacarpal and 12 months at the distal radius.

Joint laxity

The articulating joints of a foal's leg have more laxity than mature horses. This means that where there is a medio-lateral deformity of any of the types previously described, the play in the joint will be exaggerated. Some improvement in alignment can be gained by holding the joint in better position while the foal matures and the collateral ligaments strengthen.

Epiphyseal growth plates

Foals are born with bones that are incompletely formed. In order for the long bones to lengthen there is an area close to the extremities which is capable of prolific growth and conversion into bone (ossification). This area is called the physeal region or growth plate (Fig 3.8). As the cartilage continues to be produced and becomes converted to bone the length increases. Finally, the growth plate ceases to produce cartilage. The process of cartilage conversion to bone continues until the bone is solid throughout its length. The bone has then reached its final length and will not increase for the rest of the horse's life.

The thickness of a long bone is increased in a different manner. The inner cellular layer of periosteum deposits bone on the surface to increase the thickness. Also, the medullary cavity is widened by the absorption of the bone lining it. Bone thickening continues for much longer than lengthening and is affected throughout life by environment and exercise and possibly injury.

Stimulation of growth plates

The growth plates that affect lower limb conformation are stimulated by the stress of weightbearing or compression. Growth plates respond to compressive forces in 2 ways; initially the rate of production of cartilage increases, but where stress is excessive growth slows and may stop (Fig 3.12). The foal with an angular limb deformity may suffer excess compression on one side of the growth plate

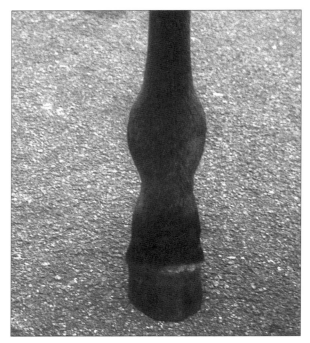

Figure 3.5:
A foal's right hind leg and foot showing very slight fetlock varus.

Figure 3.6:
Eyelining the leg in **Figure 3.5** shows the foot to be high on the medial side (right) and crushing the lateral side (left).

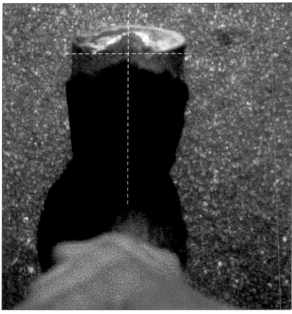

Figure 3.7:
The foot in Figures 3.5 and 3.6 is trimmed to balance to the long axis of the pastern.

damaging the cartilage and stopping growth. This perpetuates and may even worsen the angulation. The imbalanced growth across the growth plate results in an asynchronous longitudinal growth rate.

NB: ALD has causes other than an asynchronous longitudinal growth rate. Other causes include; joint laxity, defects in the development of cuboidal bones and small metacarpal bones, traumatic luxation or fracture of the carpal bones. They can all contribute to damage to the growth plate.

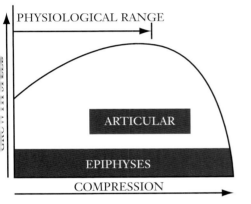

Figure 3.12: Graph depicting the response of growth plates to compression.

The process by which the growth plates are stimulated by some stress explains how many ALDs improve without assistance or intervention. We all experience the immature foal with significant carpal valgus that straightens in a matter of weeks (Figs 3.10 and 3.11). In such a case, the growth plate of the distal radius is compressed more on the lateral side, and therefore grows faster than the medial side, thus straightening the leg. As it straightens, the compression through the growth plate levels and the production of cartilage becomes more even.

Since it is the normal physiological process for growth plates to auto-correct, why do we see mature horses with distinct ALDs? Excessive compression will stop the development of bone at the growth plate, thereby consolidating an ALD. It would appear that there are 2 main reasons why compression in a growth plate may be so excessive as to stop the growth plate functioning; the foal is born with such an extreme ALD that the compression through the side of the growth plate that requires stimulation is absolute, or the foal is exercising to a degree that causes excessive compression. Other reasons for failure of the growth plate to correct itself are direct trauma to, and infection of, the growth plate.

It is probable that ALD is frequently caused by a combination of factors. An asynchronous longitudinal growth may be the result of another condition, eg an immature foal with joint laxity and a carpal valgus conformation gives rise to excessive compression on the medial side of the distal growth plate of the radius. If this compression is beyond the normal physiological range, the growth plate shuts down on that side, the other side continues to grow.

Surgical intervention

Although not involved in surgical intervention, farriers should be aware of the basics. There are 2 types of surgical procedure for ALDs. One involves fixing, by way of screwing and wiring or stapling, the side of the growth plate that is over-developed. This allows the undeveloped side to catch up. At this point, another operation is needed to remove the stapling or screws, otherwise overcorrecting may occur. The second type of surgery is periosteal elevation (PE). A 'T' shaped incision is made in the periosteum at the site of the growth plate on the side that is not growing. The periosteum is lifted from the bone. This stimulates growth on that side. It is said that if the operation successfully stimulates the growth plate over-correction does not occur.

Biomechanical stimulation

Farriery correction of ALD is aimed at equalising compression through the growth

Figure 3.10:
A young foal with carpal valgus.

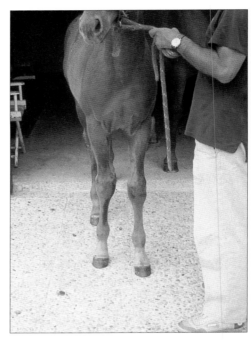

Figure 3.11:
The foal in Figure 3.10 3 months later without surgical intervention or shoeing.

Figure 3.13:
A lateral extension made with a Dalric Cuff.

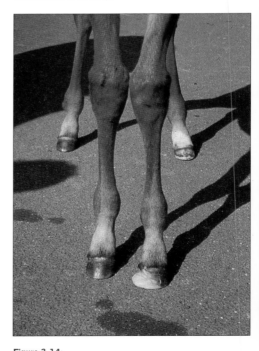

Figure 3.14:
A medial extension formed with composite.

MEDIO-LATERAL DEFORMITIES IN FOALS & YEARLINGS

Figure 3.17: Angular limb deformity (ALD); carpal valgus and fetlock varus.

Figure 3.21: An offset knee; with fetlock varus.

plate. By reducing the excessive compression on one side of the growth plate production of cartilage resumes or is increased. Note compression (Fig 3.2) does not have to be either completely equal or removed to stimulate correction; some compression is essential to improvement. Trimming the feet of foals with ALD back into balance places the foot more centrally under the limb and therefore aids the equalising of compression through the growth plate (Figs 3.6 and 3.7). A medio-lateral extension is used where trimming alone is ineffective (Figs 3.10, 3.11, 3.23 and 3.24).

Types of medio-lateral deformity

Angular limb deformity
(Figs 3.17, 3.18, 3.19 and 3.20)

This type shows a clear angulation along a medio-lateral plane in the limb. ALD may occur at the physes (growth plates) because one side of the growth plate is growing faster than the other. Without the help of radiographs to locate the exact point of angulation (eg the growth plate or carpal bones), ALD is described by joint and type, eg carpal valgus, fetlock varus. It is quite normal for there to be some slight carpal valgus in foals up to the age of weaning. Carpal valgus can be seen at the walk and standing. Eyelining will confirm that the lower leg is in alignment and therefore the deformity must be at the carpus or above. Fetlock varus is seen at the walk and by careful flexing of the fetlock joint.

Offset knee
(Figs 3.21 and 3.22)

This is seen in the area of the carpus. The radius and large metacarpal do not appear to line up through the carpus, the large metacarpal being deviated laterally. An offset knee creates uneven stresses within the limb and hoof. It is often associated with medial splints, fetlock varus and toeing-in. The hoof capsule is distorted, being flared medially and upright or under-run laterally. Wear is on the lateral solar surface of the foot with break-over at the lateral toe quarter.

CHAPTER 3

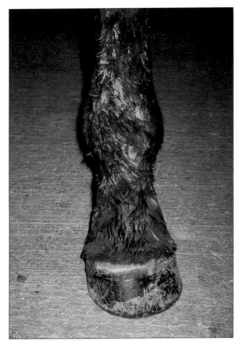

Figure 3.15:
A foal showing slight varal deformity of pastern and hoof capsule. There is not enough horn to trim into alignment.

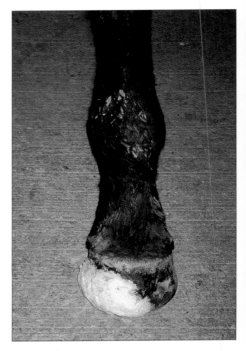

Figure 3.16:
Foal (Fig 3.15) with a small composite extension applied.

Figure 3.18:
Epiphysitis may be caused by uneven weight distribution or an unbalanced diet.

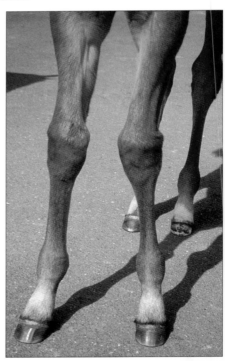

Figure 3.19:
Angular limb deformity (ALD) seen as carpal valgus (left fore).

28　FARRIERY – FOAL TO RACEHORSE

MEDIO-LATERAL DEFORMITIES IN FOALS & YEARLINGS

Figure 3.24: Rotational deformity.

Rotational deformity
(Figs 3.23 and 3.24)

This is a rotation along a horizontal plane of part or all of the limb. It is mostly seen in the front limbs, causing a severe turning out. It is best observed at the walk. Care should be taken not to confuse it with carpal valgus. The foal will dish in a similar manner to one with an ALD at the knee, observation of the foal standing will confirm the type of deformity. Quite often the elbow is tighter into the chest on that limb. The hoof capsule is usually distorted, being flared laterally and under-run medially, with the medial bulb and heel shunted up. Wear and/or compression occurs on the medial solar surface of the foot with break-over at the medial toe quarter. A horizontal rotation may be combined with or causing an ALD. In the author's opinion, at least 50% of horizontal limb rotation deformities improve by about 12 months of age.

Farrier treatment of medio-lateral deformities

Angular limb deformity

In most minor deformities, some uneven wear and slight distortion of the hoof capsule can be corrected by trimming the foot to a 90° plane to the long axis of the pastern up to the age of 12 weeks (Figs 3.6, 3.7 and 3.25). After 12 weeks the foot should be trimmed at 90° to the cannon (Fig 3.8). Trimming every 2 weeks where there is a clear though slight deformity is indicated.

In severe cases, where trimming alone is ineffective or where after trimming, the foot does not fully contact the ground, a medial or lateral extension is recommended (Figs 3.26–3.31). Shoes of any type on foals are likely to restrict hoof capsule development (Figs 3.28 and 3.29). An extension made of composite material (Atlantic Equine, Rugby, UK; Innovative Animal Products, Minnesota, USA) does not restrict the foot and has the added advantage of being able to be trimmed at a later date. Two methods of making medio-lateral extensions are described in Chapter 4.

Offset knees

Offset knees cannot be improved by trimming or shoeing; however, their secondary effects, ie splints, fetlock varus, toe-in/hoof distortion, can be reduced by sympathetic trimming and/or shoeing. Trimming should aim at

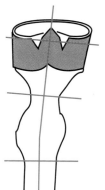

Figure 3.25: Trimming the foot to a 90° plane to the long axis of the pastern.

CHAPTER 3

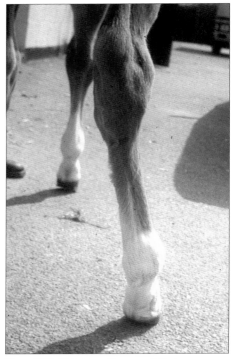

Figure 3.20: ALD deformity seen as fetlock varus.

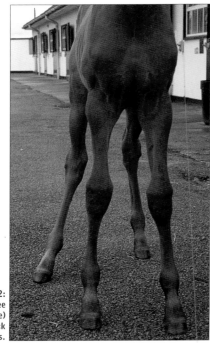

Figure 3.22: Offset knee (right fore) causing fetlock varus.

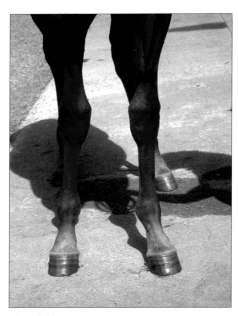

Figure 3.23: Rotational deformity (left fore).

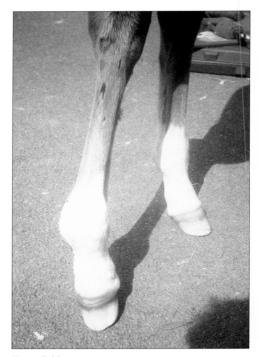

Figure 3.26: Foal with fetlock varus (Fig 3.20) flexing the fetlock showing severe ALD.

removing the excess growth and flare on the medial aspect and solar surface of the foot. The solar plane should be at 90° to the long axis of the large metacarpal.

Rotational deformities

Rotational deformities are perhaps the most difficult to trim or shoe. Trimming to a longitudinal balance and reshaping the foot will reduce hoof capsule distortion. Many rotational cases improve spontaneously. Although farriery appears to play no part in this improvement, it is important to maintain a well balanced foot to reduce uneven stresses that may adversely affect the limb. In the author's experience, aggressive over-corrective trimming of rotational deformities, (ie reducing the lateral solar aspect) does not improve these cases. The hoof capsule distorts by the medial bulb and heel shunting more, and the hoof capsule rotating medially while the limb remains rotated outward.

Summary of personal observations

Most cases of ALD associated with immaturity spontaneously correct at an early age. Toeing-in is usually the result of a combination of offset knees and fetlock varus. The normal development of a foal's limbs means that a healthy, developing foal may be slightly carpal valgus at weaning. Hind fetlocks are often varus. Most improve with sympathetic trimming. Over-corrective trimming does not improve an ALD. It will almost certainly radically distort the hoof capsule. Medio-lateral extension shoes work when they reposition the foot under the limb. In order for them to work they must be applied to a correctly prepared foot. ALD is unlikely to be improved in: PI and PII after 3 months, the distal MIII after 6 months, and distal radius after 12 months. In cases of ALD, if improvement is not seen in 2 weeks, it is unlikely to occur without intervention. The most severe carpal valgus, at over 3 months, is often seen in conjunction with a limb rotation. Carpal varus (Fig 3.33), fetlock valgus (Fig 3.34) and rotation combined with an offset knee are all rare, but they do occur.

Figure 3.27:
Foal (Figs 3.20 and 3.26) with lateral extension applied.

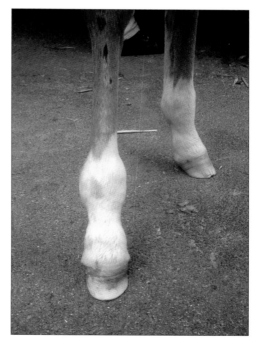

Figure 3.28: Foal
(Figs 3.20, 3.26 and 3.27) after 6 weeks; 2 lateral extensions were applied in this time.

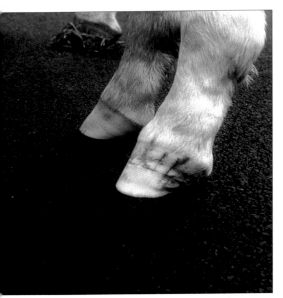

Figure 3.29:
Restriction to hoof capsule by a cuff type glue-on shoe causing some coronary band lesions.

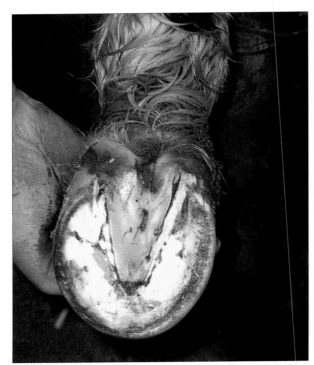

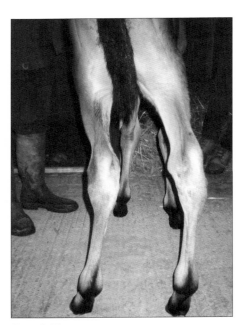

Figure 3.32:
Windswept conformation showing slight valgal deformity (left hind) and severe tarsal and fetlock varus deformity (right hind).

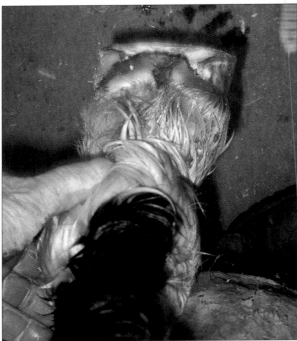

Figures 3.30 and 3.31:
Six months later the foal (Figs 3.26–3.29) has a normal hoof capsule and good alignment.

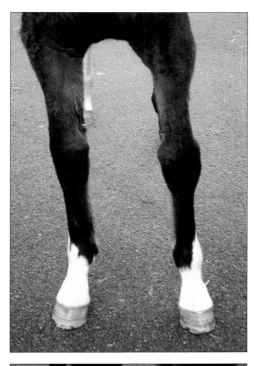

Figure 3.33:
Carpal varus deformity (bow-legged). This is uncommon – offset knees are often incorrectly described as bow-legged.

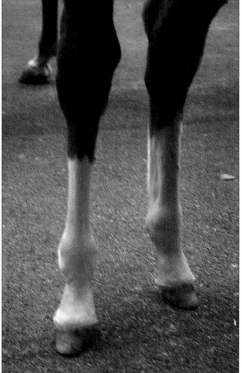

Figure 3.34:
A valgal deformity from the fetlock distal. True valgal deformity including the fetlock distal is very rare. Foals with a rotational deformity (Fig 3.23) are often incorrectly referred to as fetlock valgus.

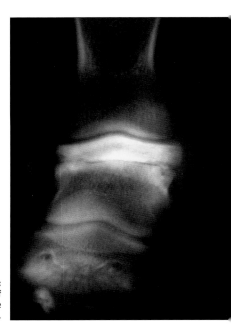

Figure 3.35:
Dorso-palmar radiograph of Figure 3.34 showing the severity of deformity.

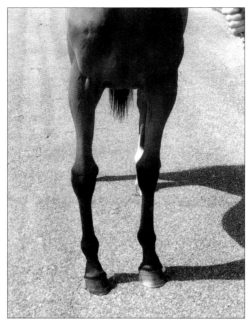

Figure 3.36: An offset knee with varal deformity of both the carpus and fetlock. The prognosis for such cases is extremely poor.

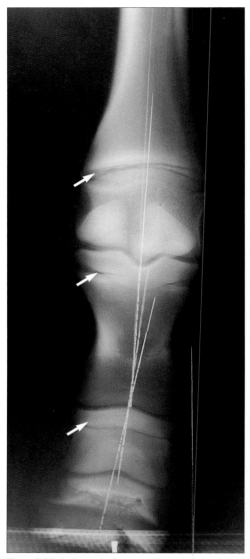

Figure 3.37: Radiograph of one-month-old foal showing an ALD deformity and open growth plates (arrows).

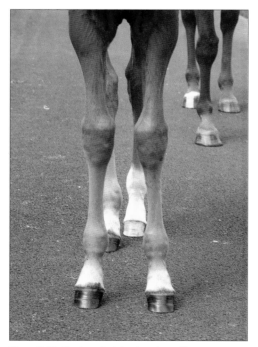

Figure 3.38: Good medio-lateral conformation.

CHAPTER 4

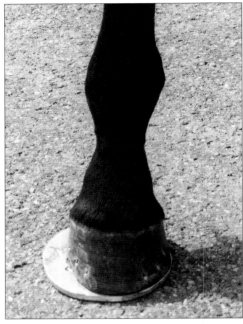

Figure 4.1:
The offset knee and varal deformity (Fig 3.36) with aluminium nail on extension applied.

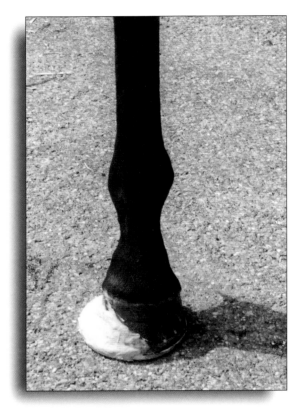

Figure 4.2:
The extension has been covered with composite material to prevent loss.

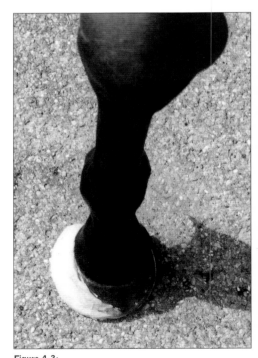

Figure 4.3:
The lateral extension seen from above.

Two methods of creating medio-lateral extensions for foals and yearlings

Method 1:
Aluminium lateral extension shoe

This was the first type of medio-lateral extension shoe that the author used. Although glue-on extensions (Baby Glu, Mustad Hoofcare SA, Bulle, Switzerland; Dalric, Dallmer, Salzhausen-Putensen, Germany) and composite extensions (see below) have largely surpassed the need for nail-on extensions, there are occasions for their use. They give a more rigid base across the ground surface of the foot, thereby reducing capsular distortion. There is also more resistance to wear than with a composite extension.

Materials and equipment
Aluminium plate $3/16"-1/4"$ (5-7 mm)
Band saw with blade for aluminium
Composite filler
Drill with $1/8"$ (3 mm) bit
Shoeing tools
Fuller/creaser

Method
Trim the foot to the balance appropriate to the condition being treated. Hold the plate on the foot and mark around the distal border with marker. Add the required extension and cut out. Mark the nail-hole position and drill. The fullering can be added after initial drilling and the nail-holes run through again. The foal or yearling may require sedation (eg Domosedan, Pfizer Ltd, Kent, UK; Torborgesic, Willow Francis Veterinary, West Sussex, UK) and the shoe nailed on. The nails are clenched. The composite is applied above the extension and smoothed to reduce the chance of the shoe being pulled off (Figs 4.1–4.3).

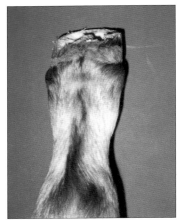

Figure 4.4:
Composite extension. The foot is prepared and cleaned with acetone. The side that will receive the extension (left) is trimmed down and cleaned. The other side (right) is untrimmed so that after the composite hardens, the whole solar surface can be trimmed to balance.

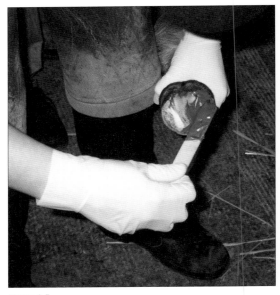

Figure 4.5:
The composite is applied and reinforced with fibre-glass cloth.

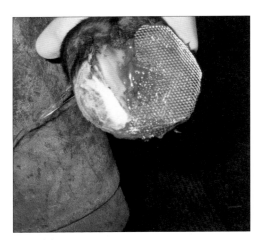

Figure 4.6:
Aluminium mesh cut to the final extension size is embedded in the composite and covered.

Figure 4.7:
The finished composite extension is rasped to the correct medio-lateral balance.

Method 2:
Composite lateral extension

This type of medio-lateral extension was first used and initiated by the author in 1992. Although glue-on extensions shoes reduce nailing risk, they require a great deal of time, skill and special equipment. The big advantage of the composite extension is its lack of restriction of the hoof capsule. Also, it can be trimmed at a later date to continue its biomechanical effect without the time and cost of a new extension. Foals and yearlings always require sedation with this method.

Materials and equipment
Acetone
Aluminium gauze
Composite (Equilox II)
Fibre-glass cloth
Milar wrap 200 mm × 200 mm
Mixing tubs
Scissors
Tongue depressors

Method
Trim and clean away all loose horn on the solar surface of the extension side. Rasp the outer hoof wall on the same side. Only trim down the wall on the side that the extension will be attached (this will result in a temporary imbalance). Cut the cloth to appropriate size. Cut the aluminium gauze to the final extension size and bend down the edge. Check the aluminium gauze for correct size. Clean the foot with the acetone.

At this point the foal/yearling will require sedation. Mix the composite and apply a layer along the sole, the lateral sulci and up the wall. Take care not to apply the composite to the coronary band and hairline. Place a layer of cloth on the composite and add more composite. Work in the composite and add more cloth. Repeat and then add the aluminium gauze. Add the final layer of composite and work the whole into the desired extension. Cover with Milar wrap and hold up while curing. Once cured the extension can be trimmed to balance and tidied (Figs 4.4–4.7)

Figure 4.8:
Mustad baby-glu extension is an alternative glue-on extension that comes in a 'kit' form. The foot on the right has a toe/medial extension.

Figure 4.9:
An extension made by welding a steel extension to a $^5/_8$" x $^1/_4$" shoe. This is a very strong extension for a yearling. It has the disadvantages of weight and nailing.

TWO METHODS OF CREATING MEDIO-LATERAL EXTENSIONS FOR FOALS AND YEARLINGS

CHAPTER 5

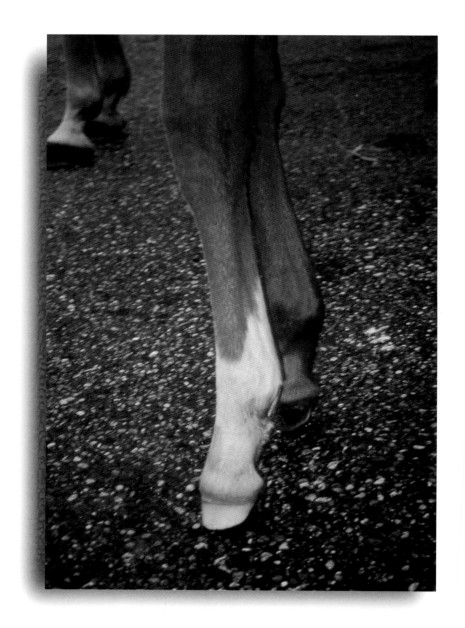

Flexural deformities in the young horse

Flexural deformities (FDs) are the group of deformities that can be seen when viewed from the side of the horse. In the young horse they are mainly divided into those that are caused by a relative contraction of the deep digital flexor tendon (DDFT) and superficial digital flexor tendon (SDFT) and hypoflexion tendons (flaccid tendons).

For many years farriery has been used as a treatment for FDs in foals and yearlings: 'knuckling over at the coronet and fetlock' described by Dollar (1898) (Fig 5.1). Historically, it is possible that the causes were not understood (are they today?) and that FDs were seen as purely a mechanical condition to be overcome mechanically (ie by trimming and shoeing).

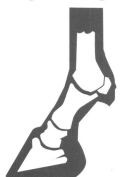

Figure 5.2: Normal (straight) hoof-pastern axis (HPA).

Contracture of the deep digital flexor tendon

There is often confusion between the hoof-pastern axis (HPA) and hoof angle. The HPA is the alignment of all three phalanges and the (dorsal) hoof wall (Figs 5.2 and 5.3). It is normal for a foal to have an HPA angle of 60° or even 65° to the horizontal. This drops to approximately 50–55° in the mature horse. A high angle (60° or more) does not necessarily mean that the foal has an acquired flexural deformity (AFD) or club foot. When contracture of the deep digital flexor tendon (DDFT) occurs the coffin joint is flexed and the HPA is broken-forward (Figs 5.4, 5.5, 5.6 and 5.7). In some cases contraction occurs so rapidly and severely that the dorsal hoof wall and pastern become vertical with the foal standing on the tips of its toes (Figs 5.8 and 5.9). This condition is described as ballerina syndrome. Hind legs are so rarely affected by an AFD causing DDFT contracture that it could be said to be a purely forelimb condition.

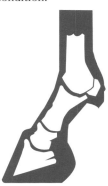

Figure 5.4: Grade I club foot.

Characteristic signs of AFD involving the DDFT only are:
- A stilted gait;
- Standing on the toe only – heel clear of the ground;
- Excessive heel;
- broken-forward hoof-pastern angle (HPA);
- Excessive wear at toe;
- Concave dorsal wall;

Club foot is the final manifestation of an acquired flexor deformity (AFD) involving relative contraction of the DDFT. It must be

CHAPTER 5

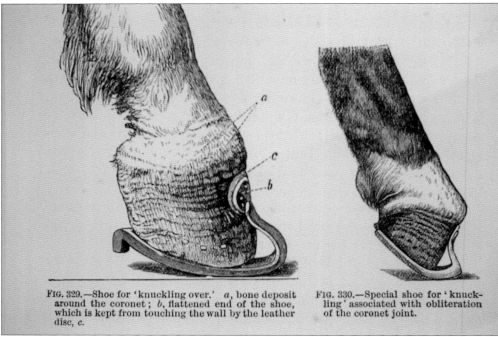

FIG. 329.—Shoe for 'knuckling over.' *a*, bone deposit around the coronet; *b*, flattened end of the shoe, which is kept from touching the wall by the leather disc, *c*.

FIG. 330.—Special shoe for 'knuckling' associated with obliteration of the coronet joint.

Figure 5.1:
Illustration from Dollar and Wheatley (1891) showing toe extension shoes for flexural deformities.

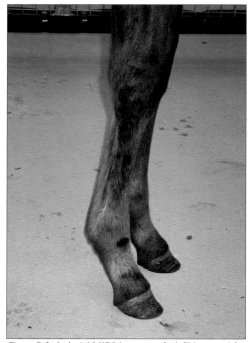

Figure 5.3: A straight HPA in a young foal. It is normal for foals to have a higher HPA angle than mature horses.

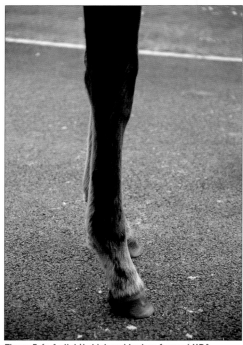

Figure 5.6: A slightly high and broken-forward HPA (Grade I club foot).

clearly understood that club foot is secondary to the condition and is not the cause. A club foot has similarities to a chronic laminitic foot and presents the farrier with many of the same biomechanical problems.

The aim of farriery is to return the foot to a normal alignment and shape in order that the horse does not have a chronic club foot which is a permanent source of lameness. The characteristic signs of a club foot are:

Lateral view

- HPA broken-forward.
- Dorsal wall angle at 65° or greater.
- Concave dorsal wall (not 'foal foot').
- Heels growing faster than toe.
- Coronary band approaching a horizontal position.
- Growth rings diverging at the heels.

- Remodelling and lytic changes including the dorsodistal border of the distal phalanx.

Grades of club foot

The Redden System of grading club feet from I to IV is as follows:

Grade I – The feet appear mismatched with the hoof angle of the affected foot 3-5° greater than the opposing foot. There is a characteristic fullness at the coronary band due to slight luxation of PII and PIII (Figs 5.4 and 5.6).

Grade II – The angle of the hoof is approximately 5-8° greater than the opposing foot. Growth rings are wider at the heel than at the toe.

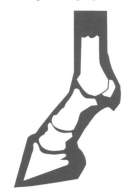

5.5: Grade II club foot.

Figure 5.8: Ballerina syndrome.

Solar view

- Foot narrower than normal (than other foot).
- Solar view shape more oval than round.
- Lateral clefts more vaulted than normal.
- Sole forward of frog apex flat or convex.
- Separation of white line/poor hoof quality at toe.

Radiographic appearance

- Lateral radiograph shows distal phalanx rotation in relation to hoof capsule and middle.

The sole will be touching the ground and the bulbs of the heel appear thickened. The heel of the foot will not touch the ground when trimmed to a normal length (Figs 5.5 and 5.7).

Grade III – The hoof wall is dished (concave) and the growth rings are twice as wide at the heel as at the toe. The impression of PIII on the sole can be seen clearly just forward of the apex of the frog. The sole shows signs of direct weightbearing and it

CHAPTER 5

Figure 5.7: The heels do not touch the ground when the foot has been trimmed (Grade II club foot).

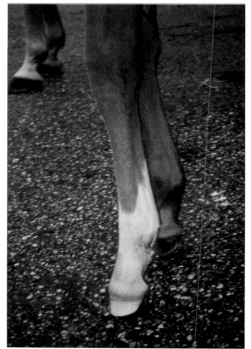

Figure 5.9: Ballerina syndrome – where the foal stands and walks on tip toe.

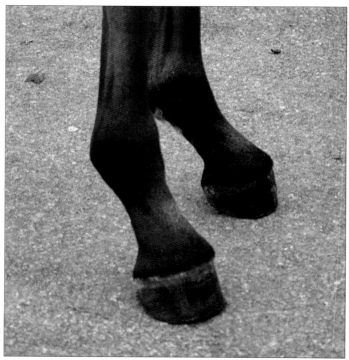

Figure 5.11: The HPA is severely broken-forward with a dorsal wall angle of above 80° to the ground (Grade III club foot).

will be bruised in many cases. The coronary band protrudes well out over the face of the anterior wall and the anterior face of PI is in alignment with the face of the anterior hoof wall. Radiographs reveal lipping and demineralisation on the distal margin of PIII (Figs 5.10 and 5.11).

Grade IV – The hoof wall angle is 80° or more (Fig 5.11) and may be very concave depending on the quality and thickness of horn. The coronary band at the heel is level or above

farrier to alter the HPA is used in the treatment of dorsal-palmar imbalances in the mature horse, eg long-toe/low-heel syndrome. It follows that, when an AFD contracture has created a broken-forward HPA, then heel trimming is indicated to achieve correct alignment. In the least severe cases (Grade I) heel trimming allows the hoof capsule and PIII to 'de-rotate'. In more severe cases (Grades II-IV) especially where the heels are already not in contact with the ground (Fig 5.5), then tendon tension maintains the HPA in a broken-forward misalignment. There are 5 possible ways of treating DDFT contracture.

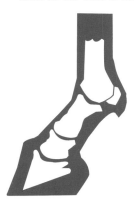

Figure 5.10: Grade III club foot.

Figure 5.12 Grade IV club foot.

the coronary band at the toe. The sole supports the weight of the bone column as it is well below the ground surface of the wall. Radiographically, the dorsodistal margin of PIII will appear rounded due to extensive demineralisation and may show several degrees of rotation (Fig 5.12).

Farriery treatment for DDFT contracture

Farriery is aimed at creating a straight HPA. In horses with normal joint flexion reducing the heels will raise the angle of the pastern and a broken-back HPA is created. Raising the heels will lower the pastern angle and a broken-forward HPA is created. The ability of the

Mechanically forcing the hoof and phalanges into alignment by means of hoof trimming, alone or in conjunction with a shoe

Where there is a broken-forward HPA and the heels are in contact with the ground (Grade III, Fig 5.10) or the heels are just off the ground (Grade II, Fig 5.5), radical heel trimming is often used in the belief that the toe will act as a fulcrum and the body weight of the foal will force the foot and/or HPA into alignment. However, it does not seem logical to believe that, when the tendon contracture has been strong enough to lift the heels off the ground despite the foal's body weight, the process will be reversed by increasing the gap between ground and heel by trimming the heels. The increased tendon tension must be transmitted

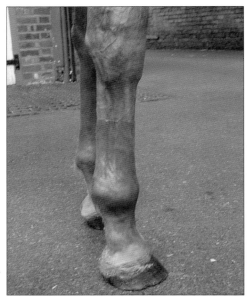

Figure 5.15:
Flexural contraction of the SDFT. The foot remains in a normal position and the pastern becomes upright. The fetlock usually knuckles forward at the walk.

Figure 5.18:
SDFT deformity after superior check ligament desmotomy – still knuckles forward.

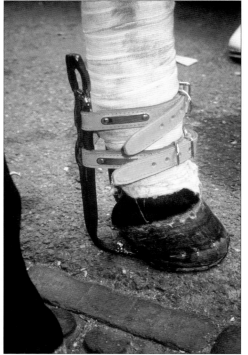

Figure 5.19: A dog collar shoe to pull the pastern and fetlock back into position.

Figure 5.13: The swan necked shoe.

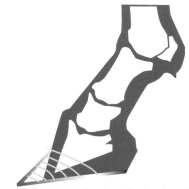

Figure 5.14: An aluminium toe extension shoe attached with synthetic material.

to the dorsal hoof wall laminae and extra stress placed upon the distal dorsal border of PIII.

Toe extension shoes have been historically used to apply additional mechanical pressure to gain realignment of the HPA. The swan necked shoe (Fig 5.13) is traditionally made of steel. This shoe has a toe extension that projects horizontally and dorsally and then curls dorsally and caudally until it rests against the dorsal wall. The rationale for the swan neck is to transfer some of the leverage stress from the solar toe area onto the more robust dorsal wall. This also lessens the chance of leverage forces lifting the shoe's heels. The authors method is a modern version of the swan necked shoe. A detailed description is in the following Chapter 6 (Fig 5.14). This shoe has the advantages of being lighter, easier to make and the composite material spreads the stress across a greater area. Another alternative is one of the glue-on shoes now available. Present day glue-on shoes are also very light and have the advantage of avoiding the need to nail to a small and possibly damaged hoof. They do have the disadvantage of being very constricting.

It is the author's opinion that the toe extension has its place in the treatment of AFD if not used aggressively. If the heels are trimmed radically and a toe extension is fitted then the additional leverage will increase the tension upon the DDFT. This additional tension must be borne within the hoof and may well result in extra damage to the dorsal laminae and remodelling of PIII. Even where the hoof capsule is forced to the ground the continual tendon pull may be enough to maintain the rotational misalignment of the phalanges. In such a case PIII is even more rotated within the hoof capsule. It is ineffective in treating the primary problem, ie the tendon contracture. The use of this shoe alone, without nutritional and especially exercise reduction, increases the risk of creating the secondary problems mentioned above.

Protecting the hoof capsule and/or the third phalanx (PIII) from remodelling changes

In cases of DDFT contracture weight is borne at the toe. There is often excessive wear with the possibility of damage to the underlying sensitive tissues. The laminae joining the dorsal wall to the dorsal aspect of PIII may already be weakened due to laminal tearing caused by PIII, leading to remodelling, lytic changes and crushing of both the papillae of the white line and the sensitive sole.

A nail-on aluminium shoe offers protection but may be difficult to apply due to lack of wall on which to nail. A glue-on shoe avoids this problem. A cuff-type glue shoe (Dalmer) protects and may also disperse stress. The author's choice of treatment, in less severe cases, is to apply a hoof hardening agent daily to the hoof wall and sole at the toe. In more

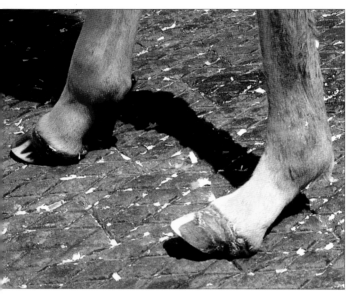

Figure 5.21:
Flaccid tendon (severe digital hyper-extension) causes a 'toe-up' conformation.

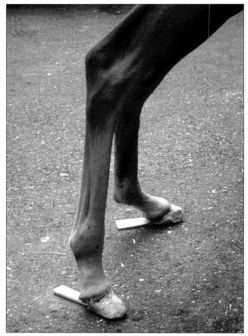

Figure 5.22:
Flaccid tendon treated with caudal extensions.

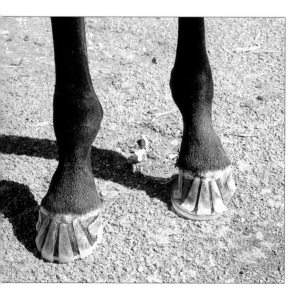

Figure 5.23:
The tab-type glue-on shoe for foals.

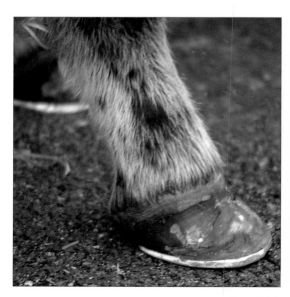

Figure 5.24:
An aluminium, nailed on toe extension with a filler around the toe.

severe cases the application of hoof repair strengthened with aluminium mesh will protect the toe pending growth.

Conservative heel trimming

In the least severe cases where the heels are seen to be in contact with the ground, conservative heel trimming should be tried. Only a small amount should be removed and the foal then observed walking and standing to ascertain the HPA and whether the heels are still in contact with the ground. If the HPA is still broken-forward and the heels are on the ground, more heel can be removed. If, however, after the initial trimming there is a gap at the heel and the foal is walking on its toe then further heel trimming is not recommended. Further heel trimming may be undertaken after 1 week of restricted exercise if the heel is then on the ground.

The most effective treatment for AFD of this type can be summed up as immediate and conservative. The best results occur on stud farms where there is close and continual observation of stock, and where action is taken at the slightest signs (stilted gait, broken-forward HPA). Such action includes a reduction/alteration in nutrition, box rest and not radically trimming the heels until relaxation of the DDFT has occurred. Progress can then be made towards returning the foot to alignment with the limb. Where this pattern is followed the incidence of chronic club foot is drastically reduced.

Relaxing the flexor tendons (FTs) by raising (wedging) the heels of the foot, thereby allowing the foot to be trimmed to a normal shape without causing more tendon tension

It has been suggested in recent years that raising (wedging) of the heels of a foal with DDFT contracture may be beneficial. There are a number of reasons given for this treatment which is contrary to many popular beliefs. Firstly, FT contracture is the primary condition. The increased heel growth creating a club foot is in response to tendon tension.

Decreasing tension on the tendon must be achieved before the club foot can be treated. Raising the heel will achieve this and emphasis can be placed upon returning the foot and phalanges to a normal alignment and shape.

Elevating the heels also allows heel trimming to take place without increasing tendon tension. Normal, even pressure may be achieved across the whole of the solar surface of the foot. Remodelling of the PIII and laminal tearing should therefore be reduced or eliminated. The author has had insufficient personal experience of this technique but the theories behind it would seem to be logical. Raising the heels in cases of superficial digital flexor tendon (SDFT) contracture have long been advocated (Dollar 1898).

Farriery in conjunction with a check ligament desmotomy

Where cases have been unresponsive to conservative treatments (farriery, box rest, splinting) or cases are presented in which rapid correction is necessary to prevent permanent changes to the hoof capsule, joints and coffin bone. Check ligament desmotomy is undertaken. The benefit of this operation is that HPA alignment is achieved immediately in almost all cases. The check ligament returns to full strength in about 6 weeks (where the foal is 6 months or less). At up to 12 months of age it is a very effective operation.

In conjunction with a distal check ligament desmotomy, corrective farriery is essential. When a distal check ligament desmotomy has been carried out, regardless of the stage of the condition, the foot should be trimmed to allow a straight HPA. Thus, in the case of a club foot the heels should be radically reduced. It is not usually necessary to use a toe extension to ensure correct HPA alignment.

Box rest

Curtailing any exercises by completely restricting the foal (and mare) to the stable is,

CHAPTER 5

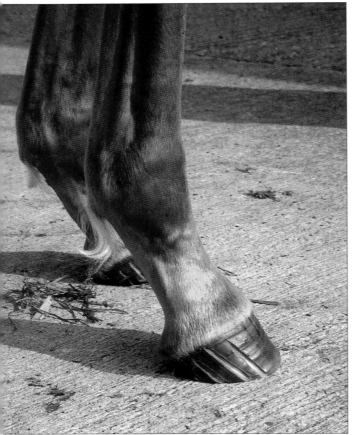

Figure 5.25:
A foal with a constricted foot grooved to promote expansion.

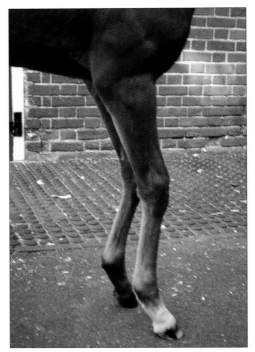

Figure 5.26:
'Preying Mantis' conformation is not helped by farriery but usually improves in time.

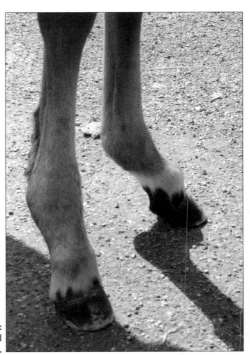

Figure 5.27:
Sub luxation of the proximal inter-phalangeal (pastern) joint.

in the author's opinion, an extremely effective method. If this is done then heel trimming can be more severe.

Discussion

The hoof capsule is very susceptible to changes in shape due to stresses from within and uneven weightbearing. Temporary changes in shape may also occur due to uneven wear. Because the horny (insensitive) structures of the young horse are more flexible than those of a mature horse and are growing faster, they are more easily affected by weight and stress factors. It is important to recognise that a club foot is the result of AFD and not the cause. Correction of the underlying condition must be successful for there to be any chance of preventing the creation of a club foot or returning the foot to something approaching normal.

The aims and possibilities of corrective farriery have become confused. In the least severe cases of AFD the foot may be the only sign of the condition. With only slightly greater than normal tension on the DDFT the weight of the foal is thrown upon the toe. This may cause extra wear. The heel, although apparently in contact with the ground, is not taking the weight of the limb and in response the hoof wall at the heel grows more rapidly. The hoof capsule has fulfilled one of its main functions. By overcoming a limb imbalance, in this case between tendon and skeletal length, the foot is now bearing weight evenly and the tendon is not under continual stress.

It is difficult to resist the temptation, when first confronted with an AFD that has already produced extra heel growth, to trim the heels immediately. If one could be certain that the primary condition was under control then it is sensible to attempt to return the foot to a normal HPA as soon as possible. Where there is clearly a gap between the ground and the heel it makes no sense at all to trim the heel. Far from 'lowering the heel' as this is often described, the opposite is achieved. Without heel contact contraction of the foot is bound to occur. This extra DDFT tension induced, especially with exercise, may lead to rotation of PIII within the hoof capsule with all the consequences mentioned previously. It is possible that the heel may be forced into contact with the ground and therefore success claimed but the phalanges remain in the rotated (broken-forward) position.

In an ideal case the condition is recognised at an early stage, ie before the dorsal hoof wall is almost vertical or a club foot has been created. The foal is box rested and nutrition is altered/reduced. Heel rasping is minimal until the condition is stabilised. The heel is then rasped by small amounts weekly. When it is decided to undertake an inferior check

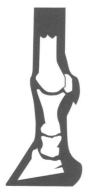

Figure 5.16: Contracture of superficial digital flexor tendon (SDFT); causing knuckling forward at the fetlock.

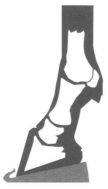

Figure 5.17: Raised heel shoe with toe extension.

CHAPTER 5

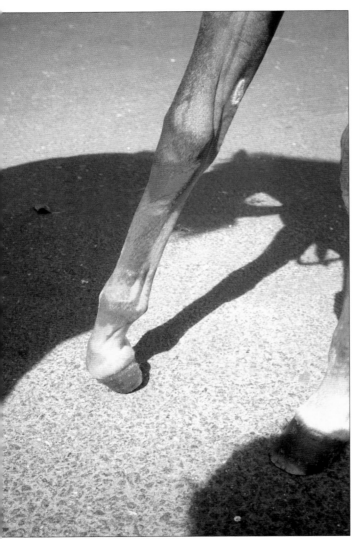

Figure 5.28: Complete contraction of the distal interphalangeal joint (DIP) (Grade IV club foot).

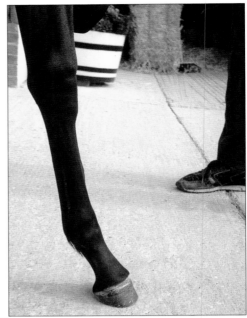

Figure 5.29:
This foal, suffering from flexural contraction, is only able to place its heels on the ground by standing in this manner.

Figure 5.20: Hypoflexed tendons; (severe digital hyperextension) causes a 'toe-up' conformation.

ligament desmotomy then the opportunity to reduce the heel radically should be taken on the same day of surgery.

Contracture of superficial digital flexor tendon (SDFT)

This condition (knuckling forward at the fetlock) occurs after 6 months of age (Fig 5.15). It can easily be differentiated from a DDFT contracture by the limb conformation. When the SDFT is involved the foot remains in normal contact with the ground and the pastern angle is raised. This creates a broken-back HPA (Fig 5.16). Traditional and present day farriery treatment is to shoe with a raised heel and toe extension (Fig 5.17). The raised heel reduces tension on the flexor tendons (allowing the fetlock to descend) and the toe extension is to prevent further knuckling.

Raising heels in cases of SDFT contracture is not a new idea (Dollar 1898). Although the author's experience is limited, in the few cases seen, there has been improvement in the pastern angle using this method. It is interesting to compare the fact that raising the heels has been an accepted treatment for many years in SDFT contracture, but only in recent years has a similar procedure been used for DDFT contracture cases. It has been suggested that the DDFT is also under excessive tension in cases when the pastern is raised and the reason that the heels remain in contact with the ground is that the route from its origin to its insertion has been shortened by the decreased angle at the fetlock. In other words, because increased tension of the SDFT has raised the fetlock the tension in the DDFT is relieved and therefore the coffin joint is not forced to flex.

Surgery for SDFT contracture (superior check ligament desmotomy) has not proved as successful as the equivalent operation for DDFT contracture (inferior check ligament desmotomy [Fig 5.18]). Various attempts have been made to force the fetlock and pastern into a more normal conformation without conclusive results (Fig 5.19).

Hypoflexion tendons

Hypoflexed tendons (severe digital hyperextension) is more commonly known as flaccid tendons. It is seen in immature foals and is characterised by the broken-back HPA with the fetlock and pastern low or even touching the ground (Fig 5.20). The foot is often weightbearing at the heels with the toe off the ground (Fig 5.21). As a result, there is no wear at the toe which becomes relatively long. The heels are worn and sometimes crushed and bruised. Most cases are mild and only require exercise for the foal to strengthen and gain a more normal conformation.

Trimming is aimed at increasing the ground surface area. Only in severe cases is a shoe recommended. If the fetlock is low or on the ground there is a danger of trauma or abrasion to it. A palmar (caudal) extension shoe will help to protect the back of the foot. It will also pull the foot back under the limb and bring the toe to the ground (Fig 5.22). Hypoflexion Tendons (severe digital hyperextension) is covered in more detail in Chapter 7.

CHAPTER 6

Making a composite and aluminium toe extension shoe

This type of toe extension shoe was first used and initiated by the author in 1994. It is a modern method of producing the same biomechanical effects as the traditional swan necked Shoe in cases of flexural deformities of the type described as contraction of the deep digital flexor tendon (DDFT) (Fig 6.1). It has several advantages over its predecessor: 1) it is lighter; 2) it requires less skill; 3) there is no nailing risk; and 4) stress is spread more evenly through the hoof capsule.

Materials and equipment
Mixing tubs
Scissors
Tongue depressors
Acetone
Aluminium (5 mm × 30 mm × approximately 150 mm)
Composite, fibre-glass cloth, Milar wrap 200 mm × 200 mm (Equilox II, Atlantic Equine, Rugby, UK; Innovative Animal Products, Minnesota, USA)

Method
Trim and clean away all loose horn on the sole and frog. Rasp the dorsal hoof wall. Cut cloth into 4 strips of approximately 25 mm × 100 mm. Cut the aluminium to required length and bend to form the toe extension (Fig 6.2). Check the aluminium toe extension for the correct size and angle. Clean the foot with the acetone. Sedate the foal/yearling. Mix the composite and apply a layer along the sole, frog and the dorsal wall. Take care not to apply the composite to the coronary band and hairline. Position the aluminium on the frog and against the dorsal wall. Apply more composite to the aluminium and hoof capsule. Place a strip of cloth on the composite and add more composite. Work in the composite and add another strip of cloth (Fig 6.3) Add the final layer of composite and cloth and work the whole into the desired extension. Cover with Milar wrap and hold up while curing. Once cured the foal can place weight on the shoe. The extension can be trimmed and tidied. The composite will heat up and then cool as it cures. This will take about 8 minutes. (Fig 6.4).

The foal must be totally restricted in exercise, ie stall rest with no paddock exercise. There is a danger that if the foal is allowed to move freely there will be excessive biomechanical leverage placed upon the dorsal wall and the laminal bond may be destroyed.

A 3–4 week restriction should be followed by removal of the toe extension for assessment. Often, normal routine may resume (Fig 6.5). Where there is seen to be improvement, the procedure may be repeated. When there is clearly no improvement, the option of inferior check ligament desmotomy should be considered.

The author has had a high level of success with the method described. Hoof capsules that, under normal circumstances, would be expected to have some chronic changes have returned to, and maintained, a normal shape (Fig 6.6). It is possible that the aluminium strip, running back along the frog, spreads the stress in a beneficial manner through the hoof capsule.

CHAPTER 6

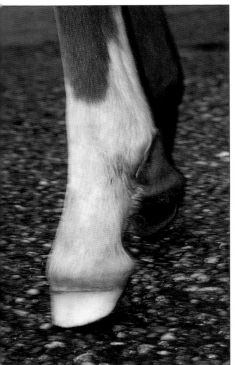

Figure 6.1: Foal with 'Ballerina Syndrome'.

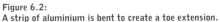

Figure 6.2:
A strip of aluminium is bent to create a toe extension.

Figure 6.3: The toe extension is attached with composite and fibre-glass cloth.

MAKING A COMPOSITE AND ALUMINIUM TOE EXTENSION SHOE

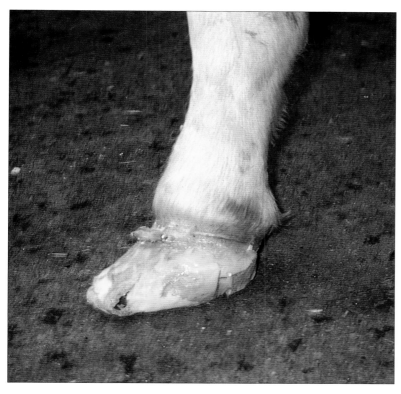

Figure 6.4: Although the finished toe extension may not be as aesthetically pleasing as the traditional 'swan necked' shoe, it has many advantages. It is lighter, spreads the stress over more of the foot and does not have the risk of nailing.

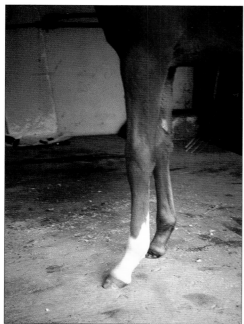

Figure 6.5: One month later the extension is removed and the foal returned to a normal routine.

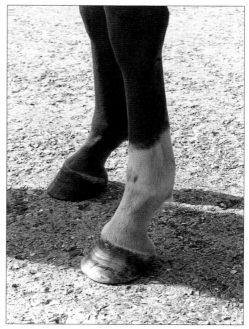

Figure 6.6: Eighteen months later the 2-year-old has a normal HPA and hoof capsule.

CHAPTER 7

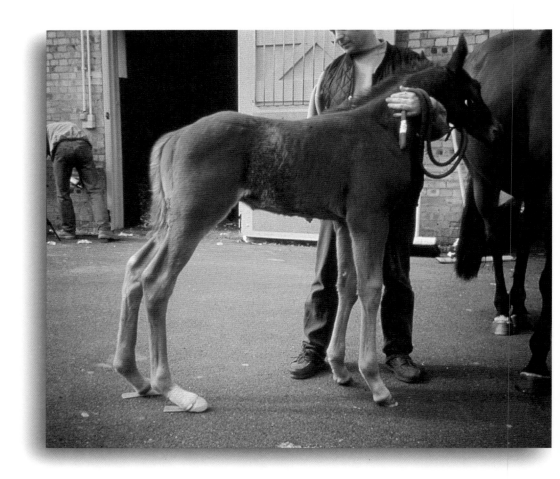

A method of treating severe digital hyperextension (flaccid tendons) in foals

Flexor tendon laxity is a relatively common limb deformity in neonatal foals. It ranges in severity and, in its most severe form, results in the foal weightbearing on the palmar/plantar aspect of the proximal phalanges and fetlock, with the toe raised (Figs 7.1 and 7.2). It is seen most frequently in premature and dysmature foals or those suffering serious systemic illness. The hind limbs are affected most commonly, although in premature foals all 4 limbs may be involved.

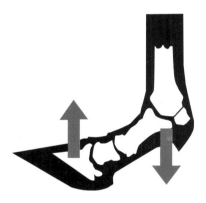

Figure 7.1: Hypoflexion tendons (severe digital hyperextension); bio-mechanical forces drive the heels forward and the toe up.

In less severe cases the condition usually resolves spontaneously, although the bulbs of the heel may need protection to prevent decubitus ulcers developing (Fig 7.3). Farriery treatment consists of trimming the heels to allow a longer base of support and to reduce rocking back on the bulbs. Some of the more severe cases fail to respond and it has been recommended that caudal extensions may be applied by taping, glueing or nailing. Caudal extension shoes work by using the foal's weight to drive the toe to the ground and hold it in the correct position during walking and standing (Figs 7.4 and 7.5).

Taping is unsatisfactory and difficult to achieve without constricting the hoof capsule. Nailed on shoes require skillful forging ability to make them and risk damage to the sensitive structures of the foot. If pulled off there is usually hoof wall damage. Glue-on shoes need to be adapted considerably to suit the individual.

This chapter describes a new technique using hoof repair composite and aluminium plates to make caudal extensions for foals with severe flexor tendon laxity.

Materials and methods

Materials and equipment

Acetone
Aluminium plate, 7 mm
Composite
Fibre glass cloth
Milar wrap
Mixing tubs
Scissors
Tongue depressors

Method

The foal should be sedated (eg Torborgesic, Willow Francis Veterinary, West Sussex, UK) and Domosedan (Pfizer Ltd, Kent, UK) to allow farriery treatment while standing. The foot is trimmed to normal proportions. The

CHAPTER 7

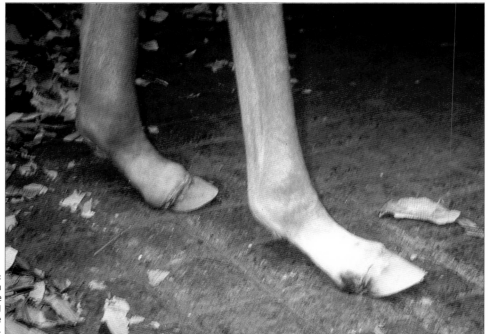

Figure 7.2:
A foal suffering from flaccid tendons. Note the lesions to the heel and pastern of the right hind.

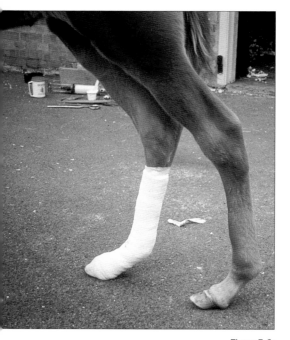

Figure 7.3:
The foal in Figure 7.2 is bandaged for decubitous ulcers. Bandaging for support is contra-indicated.

Figure 7.6:
A caudal extension made from aluminium plate.

A METHOD OF TREATING SEVERE DIGITAL HYPEREXTENSION (FLACCID TENDONS) IN FOALS

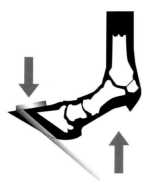

Figure 7.4: The caudal extension shoe; biomechanical forces drive the toe to the ground and bring the heels back under the limb.

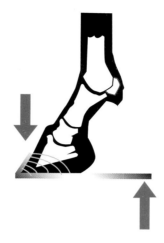

Figure 7.5: Biomechanical forces exerted upon the caudal extension hold the foot and leg in a relatively normal conformation.

sole, bars, frog and outer hoof wall are trimmed down to clean horn. The coronary band is protected with elastoplast. Any open lesions are dressed with Terramycin aerosol spray.

Caudal extension shoes are made from 7 mm aluminium plate. A strip of aluminium plate 30 mm by 150 mm is cut. This is bent 30 mm from one end to fit the angle of the toe (Fig 7.6). A layer of hoof repair composite is laid on the sole and dorsal hoof wall. The caudal extension is placed in position and another layer of composite material and strengthening cloth is wrapped around to secure it. The composite is worked to a smooth finish (Fig 7.7). Care must be taken that the lesions are not covered by any of the materials (Fig 7.8). The composite material should set prior to weightbearing (Fig 7.9).

The foal can now be exercised on a level, firm surface 4 times each day for 10 minutes. The amount of exercise can be increased progressively as improvement in both stance and at the walk is observed. Lesions to the bulbs of heel will begin to heal and the foal should be re-evaluated at a maximum of one month after the original application of the extension. The caudal extension(s) are then removed and the feet trimmed. The foal's stance and walk is then assessed to establish whether mechanical assistance is still required. If the stance and the walk are almost normal and the muscles of the hind quarters have strengthened the foal can be returned to normal paddock exercise (Figs 7.10-7.12).

Discussion

Although many cases of flexor tendon laxity resolve with careful management there are a few that fail to respond during the first few weeks of life. In order for these foals to improve they need to exercise. Many severe cases cannot exercise because of the risk of injury to fetlock and caudal areas of the pastern and hoof. If the foal remains restricted it fails to get enough exercise to strengthen the muscles of the hind quarters (semitendinosus semimembranous muscles). Caudal extensions have long been considered to be the treatment of choice in severe or non-responsive cases (Leitch 1985; Adams 1990). Previously, caudal extensions have been nailed or taped, although this method of attachment is not ideal.

CHAPTER 7

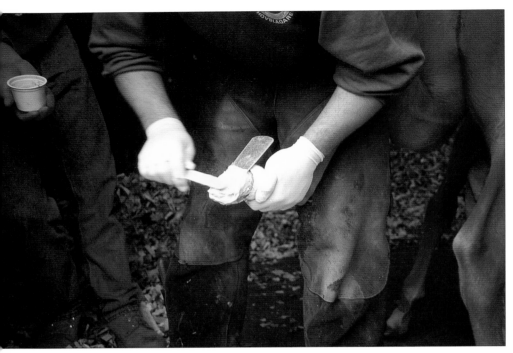

Figure 7.7:
The caudal extension is attached with composite while the foal is sedated.

Figure 7.8:
The lesions to the foot and pastern are treated and bandaged to protect them from the composite.

A METHOD OF TREATING SEVERE DIGITAL HYPEREXTENSION (FLACCID TENDONS) IN FOALS

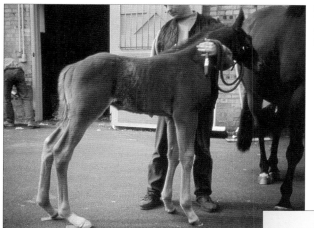

Figure 7.9:
The foal immediately after application of the caudal extension (still sedated) cannot assume a 'toe-up' stance.

Figure 7.10:
One week later, after controlled exercise, the foal assumes a near normal stance and has improved muscle development.

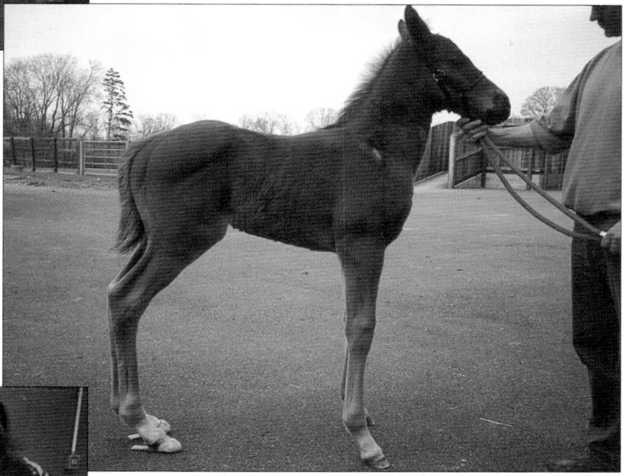

Figure 7.11:
Five weeks after the initial treatment the foal (Figs 7.2, 7.3, 7.6, 7.7, 7.8, 7.9 and 7.10) has the extensions removed and is trimmed normally.

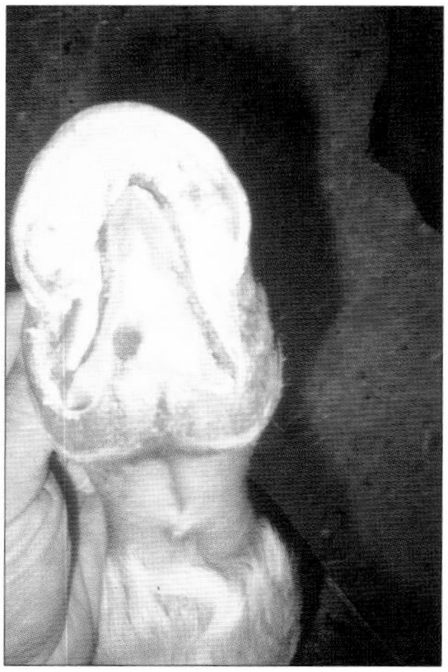

Figure 7.12:
Solar view of the foot (Fig 7.11) showing distortion due to flaccid tendons. The hoof will return to normal after 3 months.

CHAPTER 8

Preparing yearlings for the sales

Yearling sales are a very important part of the Thoroughbred industry. The farrier's preparation and presentation of yearlings prior to and during the sales can enhance the value of a yearling. There is a great deal of myth surrounding a farrier's ability to 'straighten' a yearling. The farrier cannot alter the structure of the horse. Nevertheless, there are trimming and shoeing techniques which, when applied skillfully, will cover and distract from certain faults. While this chapter is mainly about cosmetic shoeing, it should be borne in mind that good shoeing principles alone will make a horse look better and move better.

Most yearlings are not shod until approximately 6 weeks before they are sold. Few are shod behind, although it has become something of a fad in recent years to have yearlings shod behind at the sales. Some stud farms that like to give their yearlings plenty of road work do need to shoe their yearlings all round.

The usual type of shoe is steel fullered concave of $5/8"$ × $1/4"$ (15 × 7 mm) (Figs 8.1 and 8.2) section. The shoes should not have clips as these emphasise any slight deviation (Fig 8.3). The most suitable nails are ASV $1 5/8"$ or $1 3/4"$ (USA 3R or $3 1/2$ R).

The yearling should be assessed as described previously, at the walk and standing. Attention should be focused on alignment of the foot as it is placed on the ground during walking. Limbs and feet should be considered as pairs and differences noted. Trimming and shoeing is aimed at creating symmetry of stride, foot dimensions and hoof-pastern axis (HPA).

It is possible to influence the turning in or out of the foot by the preparation of the solar surface of the foot and the positioning and type of shoe. The hoof capsule of a yearling with a medio-lateral deviation will have distorted (Fig 1.19) due to the stresses resulting from uneven weightbearing. The distorted hoof capsule will exaggerate the limb deviation. Trimming the foot to a more symmetrical shape, as described in Chapters 1, 3 and 5, will improve the appearance of the limb and foot and usually improves the gait. This can be enhanced further by appropriate shoe fit (Fig 1.20).

Toeing-in

The toe-in horse is usually the result of a fetlock varus and/or offset knees. The medio-lateral balance is often (although not always) higher on the medial side. The solar surface will be stretched to the medial toe quarter and the medial wall will be more angled or flared than the lateral wall (Fig 8.4). A yearling with this conformation often paddles and may plait.

Figure 8.4: A toe-in yearling will distort the solar shape, flaring at the medial toe quarter.

CHAPTER 8

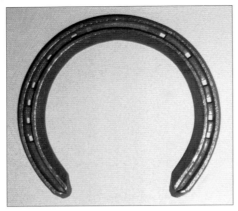

Figure 8.1:
A front yearling shoe; made from $5/8"\times 1/4"$ concave section.

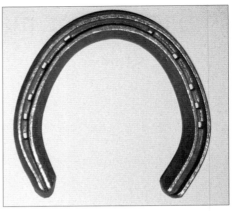

Figure 8.2:
A hind yearling shoe; made from $5/8"\times 1/4"$ concave section.

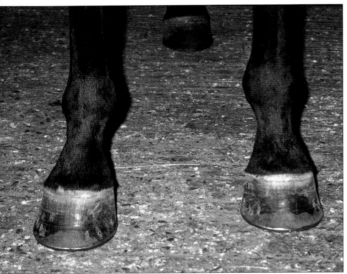

Figure 8.3:
Yearlings should not be shod with clips for the sales. It is unnecessary and may accentuate any deviation.

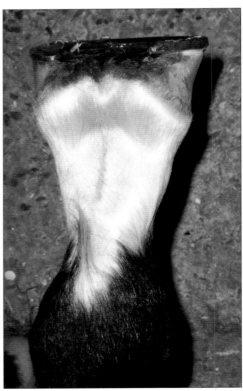

Figure 8.6:
Cosmetic shoeing for a toe-in sales yearling. The medial (right) branch is thinned and shod 'tight', the outside (left) branch is fitted full.

Foot preparation should be aimed at correcting the medio-lateral imbalance. The hoof wall should be reduced where it is higher medially. Where the medio-lateral balance is higher laterally, it should not be corrected. Medial flares should be rasped away. The symmetry of the solar surface should be trimmed so that as near as possible the hoof is equally divided either side of the frog.

Figure 8.5: A toe-in yearling; a shoe should be fitted bold and full to the lateral wall with an upright heel and tight on the medial aspect, any flare dressed back, with a pencil heel. The nails have been placed to create an illusion and the frog has been cosmetically realigned.

A shoe should be selected in a size that will allow fitting to be bold and full to the lateral wall with an upright heel and tight ('penny on a penny') on the medial aspect, with a pencil heel (Fig 8.5). The nail-holes used can further hide the true conformation and action. If the medial toe nail-hole is not used and the lateral toe nail-hole is, then the clench pattern will be biased towards the lateral. Further alterations to the shoe to change the perception of prospective buyers can be made by thinning the branch medially (Fig 8.6) and rolling the toe at 90° to the long axis of the foot.

Toeing-out

The toe-out horse is usually caused by carpal valgus or a lateral rotation of part or the whole limb (Chapter 3). The hoof capsule is distorted by flaring on the lateral aspect of the hoof wall and under-running on the medial aspect, the lateral solar aspect distorting at the white line.

The medial bulb of the heel is frequently shunted proximally.

The shoeing procedure for the toe-out yearling is basically the reverse of the toe-in therefore the guidelines in Figures 8.4, 8.5 and 8.6 can be followed swapping medial for lateral.

The solar plane should be trimmed to 90° of the long axis of the large metacarpal and the lateral flares rasped to a straight line. The solar shape should be brought back to symmetry (or as close as possible). The shoe should be fitted bold on the medial side and tight on the lateral side so that the shoe is evenly divided by an imaginary line running through the centre of the frog. Rolling the toe at 90° to the long axis of the frog will help to straighten take off. Nail-hole position can help to create an illusion of straightness by leaving out the lateral toe nail.

It may at times be necessary to trim a foot beyond balance (90° to long axis of cannon) to achieve a cosmetic improvement. To achieve the best results, walking the yearling after trimming or with just 2 unclenched nails holding on the shoe allows one to make fine adjustments.

Frog trimming

Where a foot has been trimmed and the shoe set on the foot for cosmetic purposes, the frog will still show the true alignment of the coffin bone. To achieve further cosmetic improvement the frog can be trimmed along one side to change its direction (Fig 8.5).

Club feet

The club foot is a result of a flexor tendon deformity (contraction) as a foal. The yearling may suffer from both or one (asymmetric) fore feet so deformed. Club feet vary from Grade I (discernibly different from the other foot) to Grade IV (totally knuckled forward at the pastern and fetlock with chronic changes to the distal phalanx). The shape of the hoof capsule

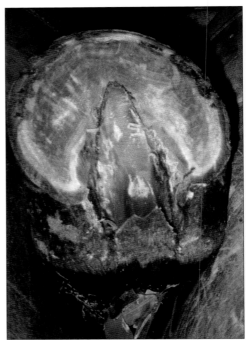

Figure 8.8:
The solar shape of Figure 8.7 shows contracted heels and an atrophied frog.

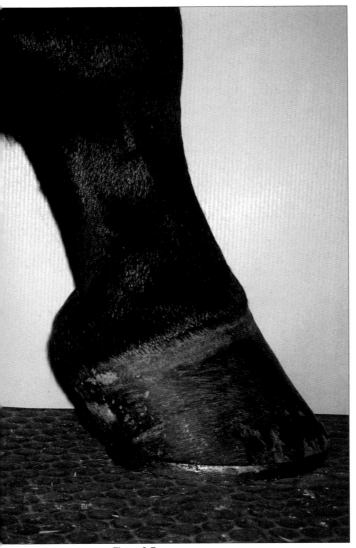

Figure 8.7:
An upright hoof-pastern axis (HPA).
Note that this is not a club foot.

accords with the severity of the condition and the subsequent treatment as a foal. Yearlings with club feet varying from Grade I to Grade III (diverging hoof rings, convex dorsal wall etc) will still be prepared for sale. The causes and grading of club feet is covered in Chapter 5.

The farrier's role at the time of sales is to present the foot and limb in the best possible light. The club foot is usually more upright than normal with the HPA broken-forward. The foot is narrower in Grades II and III than the opposite foot. Also the hoof wall is usually broken at the distal margins and the solar hoof shape is distorted. The shape resembles a hind foot, often with a flat or convex sole anterior to the apex of the frog. The foot has a close resemblance to that of a chronic laminitic. This is due to the similarity of the biomechanics involved.

The true club foot (caused by contracted tendons) has wide upright heels. Yearlings with contracted heels usually have an upright HPA (Figs 8.7 and 8.8).

Preparing the club foot for shoeing involves trimming the heels to try to create a straight HPA and shaping the solar outline towards a rounder, fuller shape (Figs 8.9–8.13). It is important to shape the shoe as full as possible and still able to nail securely. The dorsal wall should be dressed to a straight line with caution as this can make the HPA look even more broken-forward. The shoe fit should be as full to the toe as possible.

Pairing feet

Throughout this book the farrier is encouraged to treat each limb individually and to shoe each foot and limb to its needs. However, at sales, potential buyers view feet and limbs as pairs and therefore the farrier is expected to accommodate this need.

Where one foot is clearly different from another this is called asymmetry. Were the feet to be measured and compared, most people would be surprised to learn how little difference there really is between 'odd feet'. The human eye is far more accurate than we give it credit for. A pair of feet less than $1/4"$ (5 mm) different in width can be spotted 30 metres away. The eye can, nevertheless, be fooled. Judgement of the size of foot mostly comes from its width at the base. The illusion of an even pair of feet can be created using this knowledge.

The foot that is widest should be prepared first. It should be dressed normally, with just fractionally more length of wall left on, and then pulled forward and the flare dressed radically to narrow the foot. This should be done to the limit of what is still safely nailable. At this stage both front shoes should be fitted to this foot. The second narrower foot is prepared, giving attention to the correct HPA. Volume of foot is important without leaving obvious flares. It is possible in all but the most severe cases to nail the shoe that has been fitted tightly to the broad foot onto the narrower foot.

Opening heels

Opening heels is usually considered detrimental to good shoeing, but an exception can be made in some rare cases of severely contracted heels. Opening the heels is the term given to radical removal of the bars of the foot and the widening of the lateral sulci by removal of some of the heel buttress. This gives the illusion of a wider foot. Since club feet rarely have contracted heels this is not usually necessary.

Hind feet

When hind feet are shod for sales it is still best, but not essential, to shoe without clips. If the hind feet are just trimmed, then it should be remembered that the yearling may have to continue with road work and may be subject to abrasive and stony ground at the sales. The hind feet should not, therefore, be over

CHAPTER 8

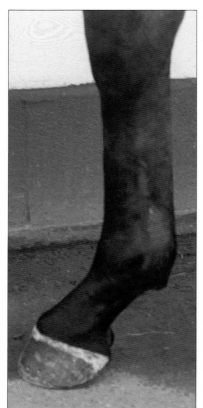

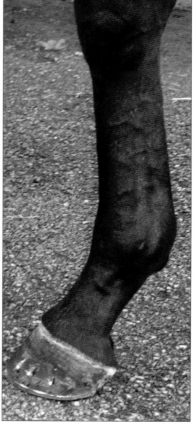

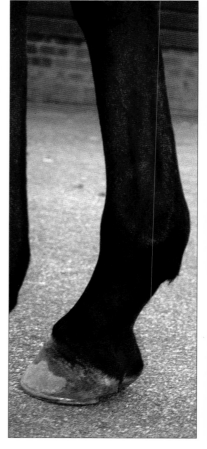

Figure 8.9:
A club foot: the HPA is broken-forward, the dorsal wall steep and the heels high.

Figure 8.10:
Cosmetic sales shoeing for club foot. The heels (Fig 8.9) are lowered and the shoe fitted full at the toe.

Figure 8.11:
Further cosmetic improvement to Fig 8.9 is gained one month later. The wall is filled to hide flares.

trimmed. The same guidelines for balance follow for the hind feet.

Back at the knee

Back at the knee is a conformational fault that is viewed seriously by potential buyers at sales. It will drop the price of a yearling considerably. Although expectations of improvement must be very guarded, some cases can be improved by judicious shoeing. The aim of trimming and shoeing is to bring the foot and lower leg back under the body of the horse.

Yearlings that are back at the knee usually have a broken-back HPA. Trimming should be aimed at correcting this. Any excess toe must be removed. Radical trimming back of the dorsal hoof wall beyond the normal guidelines is advised. The shoe should be rolled at the toe with the heels boxed to the quarters. The shoe fit should be long and wide from the quarters back, usually coming past the point of heel buttress by at least 1" (2.5 cm). Shoeing in this manner for 3 months prior to the sales often produces good results. Where the toe has been radically dressed back the toe, nails can be omitted.

Toe dragging

Yearlings often drag their toes. This is not usually a conformational problem but is just that they are weak, immature and slop along. The farrier should ensure that the foot is not overly long and if necessary fit rolled toes in front. The shoe should not be brought back under the toe of the foot as this will give the impression of an upright or club foot. The shoe should be set where the toe would have been (Figs 8.14 and 8.15). Where the problem is in the hind feet and the dragging causes excessive wear to the dorsal toe, it may be necessary to shoe. If this is the only yearling in the consignment, shoeing of the hind feet will highlight the problem. The cure is sometimes to walk the yearling more briskly to make it pick its feet up.

Hoof wall finish

Hooves that have minor cracks, old nail-holes and other blemishes can look unsightly. Rather than use a hoof repair, these can be improved by using wax. Rubbed into the superficial lesions the wax will make them disappear. Where fillers are used the hoof wall can be painted (Fig 8.12).

The hoof wall should always be finished without leaving rasp marks. These are both unsightly and detrimental to the hoof wall. Downward light strokes, following the line of the horn tubules leaves the hoof wall looking smooth. A sanding block gives an extra finish.

It is sometimes customary for the entire hoof wall to be rasped from coronary band to ground surface, taking out all minor growth rings. This may be pleasing to vendors and buyers, but it is detrimental to the hoof wall. After entering training these feet often suffer from splits and cracks as a result of this practice. If this practice has to be carried out, however, then a hoof wall treatment should be applied immediately.

Figure 8.12:
The hoof wall is painted to cover the filler.

Figure 8.13:
The solar shape of (Figs 8.9–8.12) a club foot. Note the well developed frog and wide heels. The shoe is fitted full at the toe quarters to enhance the shape.

PREPARING YEARLINGS FOR THE SALES

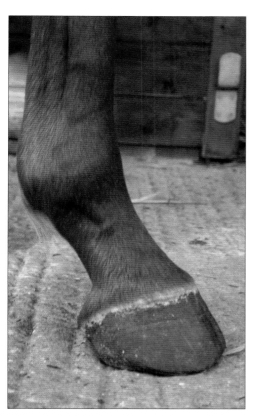

Figure 8.14:
The yearling has scraped or dragged his toes, giving a slightly club foot effect.

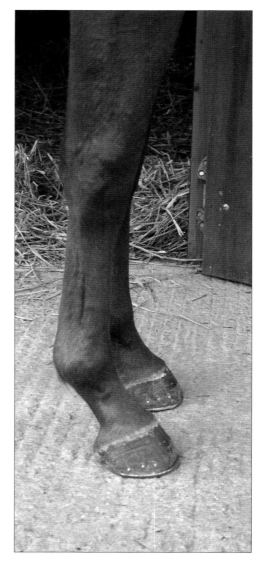

Figure 8.15:
The foot of (Fig 8.14) is shod full at the toe.

CHAPTER 9

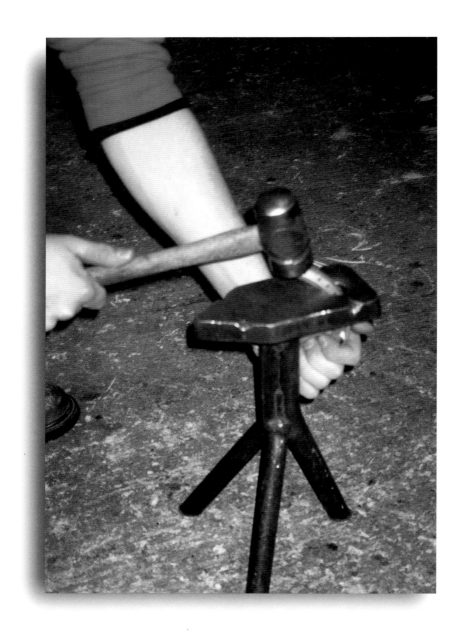

Shoeing the racehorse in training

Thoroughbreds have a justly deserved reputation for poor hoof quality and shape. Of all the breeds, farriers have the most reason to dislike shoeing the racehorse. Their hoof walls are thinner and disintegrate more easily than any other breed. The nature of racehorse training in the UK (and some other countries) means that they are subject to varying amounts of road work and are therefore trained in steel shoes (training plates), but when actually racing are fitted with aluminium shoes (racing plates; Victory Plate Co Ltd, Maryland, USA). In the UK the racehorse averages 18 sets of shoes or plates per year, far more than the average for most other types of horse. Hunters, ponies, eventers, showjumpers etc average about 8 sets per year. The frequency of shoeing has reduced in recent years with the advent of better quality plates. Most manufacturers use the strongest aluminium alloys available plus a steel toe insert for additional wear resistance. Plates now have up to two thirds the length of life of steel. Where trainers keep their horses exclusively on turf (grass), some have adopted the fashion for training their horses full time in aluminium. This practice is widespread in North America where it is rare to see racehorses on the track wearing steel.

As well as having poor quality thin hooves, the Thoroughbred is often beset with poor hoof shape. The prevention of conformation and hoof distortion have been covered in earlier chapters, problems of imbalance relating to training are covered in Chapter 10. It is in training that the farrier's ability to cope with those problems under the stress of training is truly tested. All the characteristics of good shoeing are needed to maintain soundness through a racehorse's career. The foot must be both balanced and level. The shoe must be of the appropriate size and have perfect contact with the foot. Nailing needs to be secure and high enough to be above broken horn and previous nail-holes. It is sometimes necessary to use good, sound old nail-holes due to the frequency of shoeing. Prior to the start of racing the farrier should be trying to develop good feet, saving some foot without allowing long-toes to occur.

In many yards, it is the farrier's duty to check the state of shoeing of each horse regularly. Some farriers check every horse daily before exercise. The farrier can use his own judgement as to exactly when any horse should be shod and can plan his shoeing schedule so as not to shoe a horse immediately prior to it needing plating for racing. The trainers co-operation is essential to get the best from this routine. Many trainers are able to tell their farriers a week in advance when a horse may run, while others seem incapable of deciding right up to the final one day declaration stage.

Road work

Road work is usually only seen in the UK and Ireland and this is decreasing to the the rising number of vehicles using the roads. Nonetheless, many trainers see it as an important part of the conditioning of a racehorse. Even where horses are not given conventional road work there are many racing yards that are 2 or 3 miles from the gallops.

A racehorse in road work will wear shoes at a rapid rate (Fig 9.1). It is quite usual for a racehorse to completely wear through his shoes

CHAPTER 9

Figure 9.1:
Road work; racehorses in the UK and other countries walk and trot on metalled roads to their place of exercise.

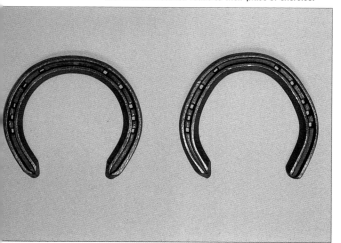

Figure 9.2:
$5/8" \times 5/16"$ (16 x 8 mm), front and hind, fullered concave training shoes; this is the most commonly used racehorse shoe in the UK.

Figure 9.3:
$5/8" \times 5/16"$ (16 x 8 mm), front, fullered concave training shoe with toe clip; most racehorses in the UK are shod with toe clips.

in 4 days, especially if young or unfit. This can cause problems of nailing and hoof loss, which continue through the season. The profile and design of shoe can increase its life. It is often better to increase width rather than thickness of shoe. Therefore a $5/8" \times 5/16"$ (16 × 8 mm) is better replaced with $3/4" \times 5/16"$ (20 × 8 mm) than with $5/8" \times 3/8"$ (16 × 9 mm) (Figs 9.2–9.5). Shoeing with a rocker toe in front and a set toe behind allows the horse to move better and removes the shoe from the area of most wear (Fig 9.6). Hard facing (welding with borium/tungsten) increases the life of a shoe although it has been suggested that this may cause excessive jarring and stress in the foot and joints.

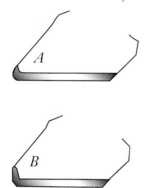

Figure 9.6: A rocker toe (A) improves break-over in front and reduces wear; a set toe (B) reduces toe dragging and protects the hind foot.

Shoe loss

There is a big difference between shoes loss that results from excessive wear and sudden shoe loss at the trot, canter or gallop. Lost shoes are one way in which the ability of a farrier is judged. In most cases this is unfair, since it is possible to shoe a horse in such a way as to guarantee that a shoe will stay on. Unfortunately, however, shoeing short and tight to achieve this will almost certainly ensure that the horse becomes unsound in the near future. Even shoes fitted 'penny on a penny' where the shoe is full and of correct size but has absolutely no overhang at the heels, will not stop shoes being pulled off.

The main causes of shoe loss are:

1. The surface exercised on. On heavy ground horses have a greater tendency to lose their action and strike the front off. Some of the surfaces used for warming up at the trot are renowned for shoe loss. Horses do not lose shoes walking or trotting on the road (Fig 9.7).

2. The rider's ability affects the horse's balance. Some exercise jockeys seem to lose shoes whichever horse they are on.

3. Pulling up rapidly after a gallop or race. Most plates that appear to be lost in a race are in fact lost after the winning post.

4. Hoof quality. Shelly hooves where the nails do not have a solid hold are more easily wrenched off.

5. The conformation of some horses means that they are always prone to shoe loss due to their action.

6. Immaturity/weakness – young or unfit horses have more of a tendency to strike off their shoes when they lose their action.

7. Farrier's ability – despite the above reasons for shoe loss the farrier can have a great effect on reducing the number of shoes lost. Since most trainers will not accept the above reasons, you might as well do everything in your power to keep shoes on.

All the rules of good shoeing reduce the likelihood of shoe loss. The foot must be prepared so that it is most suited to that horse's conformation, ie balanced. The shoe should be fitted, without extra length if it is an upright foot. Upright feet do not need extra length of shoe to support them. When shod this way, as opposed to a flat foot, the shoe heels stick out and are not covered by the bulbs of the heel. The hind toe can easily get over the protruding heels and strike them off.

The reason shoes are lost in heavy ground is not because the shoe is 'sucked off', but because break-over is delayed allowing the hind shoe to strike off the front. If the front

CHAPTER 9

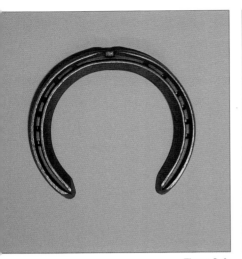

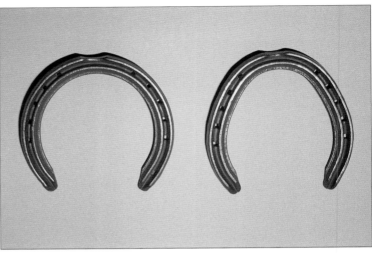

Figure 9.4:
$3/4" \times 5/16"$ (19 x 8 mm), fullered concave front training shoe with toe clip; this is a good section of shoe for feet of 5" (125 mm) or greater.

Figure 9.5:
15 x 7 DR (double runner) gives extra grip but reduced wear.

Figure 9.7: Shoe loss occurs more often in deep surfaces and especially at the trot.

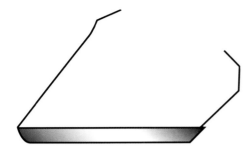

Figure 9.8: A rolled toe helps break-over and reduces over-reach and scalping injuries.

shoe is rolled (Fig 9.8) or rockered this will cause a better break-over and take the front foot away.

Nailing must be secure and well above any poor horn. Shoes are never pulled off at the toe, so the farrier should ensure that the posterior nails are still well up the wall and far enough back. It is a golden rule but there are exceptions never to nail behind the widest part of the foot. This is a good rule with justification. However, there are exceptions. If necessary a farrier can, and sometimes has to, nail posterior to the widest part of the foot in order to find sound horn. A nail in the heel quarter is worth 2 at the toe.

Clips help prevent shoe loss by stopping shoes being pushed back (Fig 9.9). All the forces exerted on the shoe are pushing the shoe back and the foot, on top of it, forward. Side clips may be more effective but have the disadvantage of taking up wall that can be utilised for nailing.

Almost without exception, it is front shoes that are pulled off. This gives lie to the belief that shoes are sucked off or drop off, as this would happen equally for the hind shoes. Because it is the hind foot and shoe that strikes off the front, careful attention must also be given to foot preparation, type of shoe and shoe fit to the hind feet. They must not be long and the shoe must be fitted well under the toe. The toe of the shoe can be rolled so that if it does strike the front heel, it is more likely to glance off.

Many textbooks say that the farrier should lower the heels of the hind foot and/or thin the heels of the shoe. This is said to delay the hind foot break-over and take off and therefore delay the action, allowing the front foot to break-over and get out of the way of the incoming hind foot. This advice regarding foot preparation and shoe type for over-reach is not indicated or logical in the author's opinion. First, altering the foot shape and breaking back the HPA in an athletic horse is playing with fire. Secondly, if we consider the argument it does not even make sense. Were the foot to be delayed in its action by this method it would have at some point to catch up otherwise the horse would split in half after a few strides. The point of catch up must be at the end of the stride – at the very point that the hind foot comes closest to the front. The whole aim must be to improve and help the horse's natural action, not to disrupt it.

Front shoes are almost always struck off on the lateral heel. This may be because the hind conformation is base narrow and/or fetlock varus. This leads the foot to grow more upright on the lateral wall and to flare on the medial. Correcting the hoof shape, the limb conformation is probably permanent, may lead to a wider action so that the front lateral heel is now missed. Careful medio-lateral balancing with a hind shoe fitted full from the lateral quarter caudally and the medial shoe branch fitted tight, enhances the action.

Where there is hoof loss at the distal margin of the wall, especially on the medial aspect, the shoe is exposed and more liable to be stepped

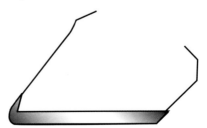

Figure 9.9: Clips help prevent shoe loss by stopping shoes being pushed back.

CHAPTER 9

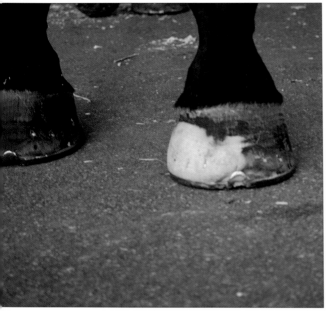

Figure 9.10:
Composite repair allows secure nailing and the maintenance of hoof shape.

Figure 9.13:
A European style front clipped aluminium racing plate with a steel wear insert at the toe.

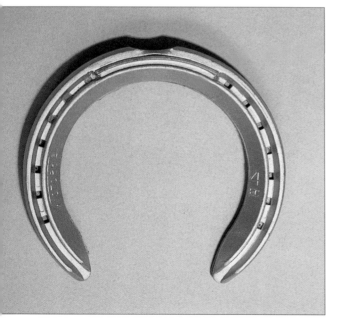

Figure 9.11:
US style 'Lite Steel' training shoes of $^5/_8$" (16 mm) section.

Figure 9.12:
US style 'Heavy Steel' training shoes of $^3/_4$" (19 mm) section.

off. This can occur on both the front or the hind feet. The shoe exposed in this area can be boxed off (the sharp corner of the section rounded) so that any tread from the opposing foot is less likely to take hold. In persistent cases some hoof filler can be used until the gap grows out (Fig 9.10).

Some horses have a habit of running the edge of one shoe down the wall of the opposing foot, against the clenches and the shoe edge, often when travelling in a horse box. This leads to nails being trodden down and shoes pulled off. This is not easy to deal with. Rounding the medial branches on the ground surface reduces the chances of their being able to get hold of the shoe. Some horses need to be bandaged on all four feet to travel.

Shoe types

Training shoes are made of mild steel which is iron with a carbon content. This gives wear and strength and has the advantage of being malleable whilst retaining its shape. The section (dimensions of shoe stock) of steel for racehorses is fullered concave (USA swedged) which combines lightness with grip (Figs 9.11 and 9.12). The concave section allows the horse to cut into and remove its foot on most ground with greatest ease. The fullering immediately fills up with grit and dirt and gives excellent grip on hard surfaces such as concrete and tarmac where a flat shoe would normally slip. The section is selected according to size of foot and work that the racehorse is doing.

Most British racehorses are shod in sections $5/8" \times 1/4"$ (15 × 7 mm), $5/8" \times 5/16"$ (15 × 8 mm) and $3/4" \times 5/16"$ (20 × 8 mm; Richard Ash Horseshoes, Somerset, UK). Light steel for racing is in either $1/2" \times 1/4"$ (12 × 7 mm), or $5/8" \times 1/4"$ (15 × 7 mm). Most yearlings are shod with $5/8" \times 1/4"$ until they begin to do more work (Figs 8.1, 9.2–9.5). Various shoe manufacturers have their own way of sizing shoes which leads to confusion. The simplest way to size any shoe is by width. Small yearlings may be a 4" (100 mm) width and large mature flat racehorses up to $5^1/2"$ (140 mm). The majority of racehorses fall between $4^1/2"$ and 5" (115 × 130 mm) (Figs 9.11–9.20).

Shoes and plates are fitted cold in racing. This means that the onus is on the farrier to trim the foot level. Shoes are adjusted on site to each individual foot. The best tool for this is a stall-jack. These vary from farrier to farrier. Many have their own preferences. The basic design is a plate of steel approximately 5" × 3" (12.5 × 7.5 cm) and $3/4"$ (2 cm) thick with a 1" (2.5 cm) hole in it. The plate is attached to a tripod (sometimes a stake in North America) at a height of about 18" (45 cm). On this the shoe can be opened, turned in and levelled with ease (Fig 9.21).

Some farriers persist in carrying about miniature anvils. These are a wonderful design for making shoes but they are of little use for cold fitting.

Although racehorse shoeing and plating is best carried out cold, it does not mean that the shoes cannot be prepared hot prior to shoeing. Ready made shoes have improved immensely during the last 20 years. Nonetheless, they still lack the finesse that can be given by some small adjustments in the forge. The more work done on shoes prior to shoeing the better the job and the easier on the farrier's back. Shoeing is hard enough on the back without having to remake the shoe under the horse.

The 'footy' horse

Many Thoroughbreds' feet are so thin walled and soled that even the most careful shoeing leaves them footsore for some time after shoeing. We should not forget that despite farriery existing for some 2,000 years, we are still banging nails into feet. Even when the nails do not penetrate (prick) the sensitive tissue they are very close to it and do cause pressure. Without depth of foot the nails enter

CHAPTER 9

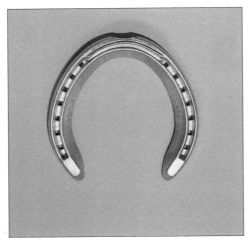

Figure 9.14:
A European style hind clipped aluminium racing plate with a steel wear insert at the toe.

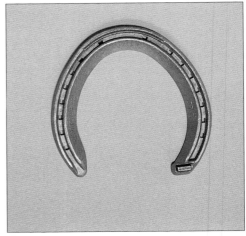

Figure 9.15:
A European style hind aluminium racing plate with an outside calk of no more than $3/8$" (10 mm).

Figure 9.16:
A 'bonded plate' with an anti-concussion rim pad.

Figure 9.17:
A hind plate with a 3° wedge.

near the laminae and run up the wall close to them.

In less severe cases there needs to be some planning so that horses are shod when sore-footedness will not affect their training. They should not be plated immediately prior to racing but possibly 4 days before.

In severe cases glue-on shoes completely remove the need for nailing. This allows the foot to flex without the restriction of nails (Fig 9.22).

Glue-on shoes

There are 2 types of glue-on shoes. The first is an aluminium plate bonded in plastic with finger-like tabs projecting up to attach to the outer wall. This is glued on using cynoacrylate (superglue). The second type is a complete cuff that is riveted on to a conventional shoe. This is then glued on with an epoxy resin.

Rim pads are bonded under the racing plate between the foot surface and the width of the section. A variety of materials are used from foam plastic to felt and sorbothane (Fig 9.23). Some manufacturers supply ready bonded plates, (Fig 9.16); otherwise platers can buy or make the rims themselves and glue them to the plate. Rims usually require riveting at the heels or the movement of the foot will spread them.

Hoof rebuild

Another way of overcoming a chronically 'footy' horse is by using the rebuild technique. This involves cleaning the wall thoroughly and applying a thick layer of composite to the outer wall. This is nailed to in the conventional manner without the nails entering the wall to the same depth (as described in Chapter 10).

Plating for dirt track racing

Introduction

Most countries race on dirt. The surfaces vary considerably from track to track. Some may be, as their name suggests, just dirt. Others are mixtures of soil or sand with synthetic substances. Others have chemicals added to prevent freezing during the winter. The aim always is to achieve a consistent and good going. All race tracks are under pressure to produce a fast surface. The common factor with all dirt tracks is the looseness of the surface. Inclement weather wil alter the going of most dirt tracks radically, making them deep or slick.

Steel training plates

Most racehorses are trained at the track. They come in from the farm or sales and unless they return home for a long lay-up, they train and race for long periods at the same track. They usually arrive at the track wearing steel. Traditionally, training plates were hand made by running a flat section of steel through a swedge block. This still forms part of the union test for some race tracks in the USA. Training plates come in 2 swedged profiles; lite steel ($5/8$" × $1/4$", 16 × 6 mm) (Fig 9.11) and heavy steel ($3/4$" × $1/4$", 19 × 6 mm) (Fig 9.12). The nails used to attach training plates are city head sizes 4RN or $3^1/_2$RN. In some instances, for larger feet shod with wide section, $4^1/_2$ or even 5 city head may be used.

Aluminium racing plates

Because they are not in abrasive conditions, most racehorses remain in aluminium racing plates for the duration of their time in training. The plater will usually shoe his trainer's horses on a 30 day schedule. Racing plates come in a wide variety of styles to suit different conditions. These include various types of heel, including; mud calks, blocked heels, stickers and calks. The plater may also alter the style of the plate as in turn-downs. The toes vary from a queen's plate which is flat with a steel wear insert to different heights of toe-grab (Fig 9.24). The nails used to attach racing plates are city head sizes 3RN or $3^1/_2$RN (Fig 9.25).

CHAPTER 9

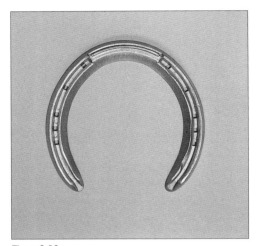

Figure 9.18:
An outer rim plate.

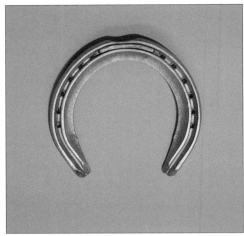

Figure 9.19:
A front racing plate made from an extruded aluminium section.

Figure 9.20:
Aluminium straight bars and eggbars are manufactured in sections suitable for racehorses.

Figure 9.21:
The stalljack is the most suitable tool for altering shoes and plates; they vary greatly in style from farrier to farrier.

Foot preparation

Because of the nature of the track surface the hoof is prepared for the plate in a manner which reduces clogging of the foot with the track material. The sole is trimmed out to a smooth convex surfaced. This is called cupping the sole. The frog is trimmed completely to juvenile horn and the heels are opened (Fig 9.26). Many platers pack the sole the evening prior to shoeing in order to soften the frog and sole (Sole Pack, Hawthorne Products Inc, Indiana, USA). Where a sole and frog are hard and bound the foot can be scorched with a gas flame to soften the horn to allow easier removal (Fig 9.27).

Feet with under-run heels are even less able to cope with dirt than turf. It is therefore essential that the heels of the plates have length to protect the caudal area of the foot.

Plating for turf

Racing plates for the turf are sold with a number of options: clipped, non-clipped or side clipped hinds (Figs 9.11–9.15). Within British Jockey Club rules, racing plates cannot have a toe-grab and can only have a calk projecting $3/8$" (10 mm) (Fig 9.13) below the ground surface on the outside of the hind plates. Nails must not be prominent.

Most racehorses run with plates all round. Occasionally, especially early in the season they may run in their shoes. Trainers may also chose to race their horses in light steel (Fig 8.1) on the hind feet and racing plates in front. Unlike the USA, trainers in the UK do not have to declare the footwear prior to racing. The weight of a set (4) of racing plates is approximately 6 oz compared to shoes (training plates) which are about 16 oz. If the old adage about an ounce on the foot being worth a pound on the back is true, then many trainers and farriers have indulged in their own private handicapping system over the years.

Usually when plating for a race, only enough horn is removed to ensure that the foot is level. Then when the horse is shod after the race, the excess foot can be removed to allow new nailing to take place. Obviously the farrier needs to use his judgement to make this system work. If the horse has not been shod for a month and may be in racing plates for three weeks, then leaving the foot overlong before plating will be harmful.

CHAPTER 9

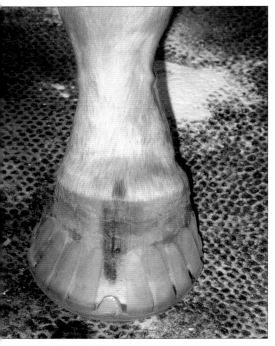

Figure 9.22:
'Race-glu' shoes are aluminium racing plates bonded inside plastic; they are glued to the outer wall.

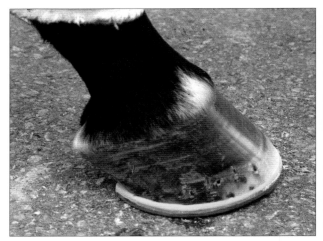

Figure 9.23:
Various anti-concussion materials have been bonded to racing plates to reduce jarring.

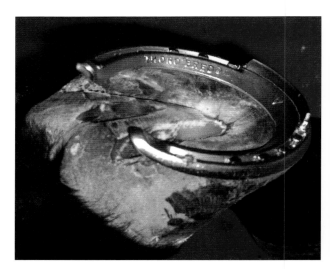

Figure 9.24:
Toe-grabs vary in height; this plate also has 'turn-downs' at the heels.

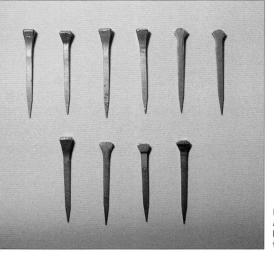

Figure 9.25:
A variety of nails is used in racing; in the US only city heads are used; in the UK, E heads are usually used in the training shoes.

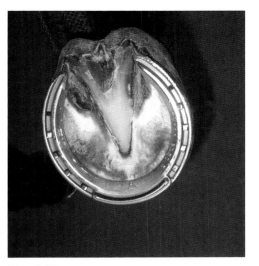

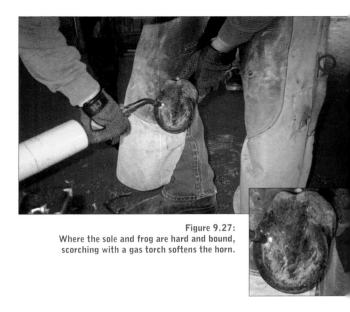

Figure 9.27:
Where the sole and frog are hard and bound, scorching with a gas torch softens the horn.

Figure 9.26:
Plating for dirt track racing; the sole and frog are well trimmed with good clearance along the sulci.

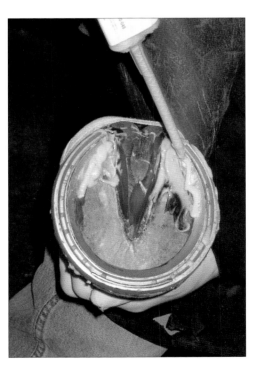

Figure 9.29:
The plate is attached to the distal wall, the outer wall caudal to the quarters, the bars and lateral sulci.

Figure 9.28:
A method for attaching an aluminium plate using acrylic adhesive, advanced by W. Champagne. The plate is thoroughly cleaned with an abrasive wheel and heated prior to gluing.

CHAPTER 10

The racehorse in training: problems associated with imbalance

Balance has been described in Chapter 1. For the purpose of this chapter, an imbalance is where the foot alignment deviates sufficiently from the ideal to cause or has the potential to lead to, lameness and other conditions which effect the performance of a racehorse.

It is often stated that if you look in the winners' enclosure you will see all sorts of crooked legged horses. This is true. Good conformation rarely made a horse run faster, but it does give rise to fewer injuries. It is difficult to win races while lame in a stall. Many racehorses will succeed in spite of all our attempts to stop them. It is also true that a great number have had injuries prevented and overcome by the recognition of imbalances and the correct application of the principles of balance.

The causes of imbalance

Congenital

The Thoroughbred has been bred to race for over 200 years. Stud farms send their brood mares to stallions who are proven on the racecourse. Some of these do not have particularly good conformation but often a great racehorse can overcome deficiencies with strength and spirit. This is a characteristic that breeders would hope to pass on anyway. Brood mares are not necessarily successful racehorses. Frequently, a mare that is unable to race through a congenital defect or injury is used for breeding due to her perceived good blood lines.

Although Thoroughbreds are not exposed to the fads that bedevil many other breeds, bred for show, it is quite possible for genetic faults to be passed on through the generations.

Congenital deformities are those which are either present at birth or show themselves soon after birth.

One fault that is often passed from the broodmare side is severe offset knees. In the author's opinion, any mare with this conformation seems to pass it on to a large proportion of her offspring. You rarely see severe offset knees in stallions because they cannot stand training.

Acquired limb deformities

Club feet (Figs 5.3–5.7) result from an unresolved contraction of the deep digital flexor tendon (DDFT) (covered in Chapter 5). We know that tendons do not contract but this term helps us to understand the biomechanics in the young foal. The result is that the DDFT pulls on it's insertion in the distal phalanx. This causes a rotation of the coffin joint, pulling the heels off the ground, tearing the dorsal laminae and forcing the distal anterior margin of the coffin bone into the sole. If not treated immediately, the condition becomes permanent. Cosmetic farriery will often disguise the condition enough for it to pass successfully through the sales. Once in training, however, the condition becomes apparent.

Angular limb deformities (ALD) occur in many foals at or after birth (covered in Chapter 3). The most common are carpal valgus (knock

Figure 10.1:
Three-quarter shoes were once very popular in British racing; in some cases every horse in a yard was shod with them, regardless of the type of interference. In some brushing cases they may help.

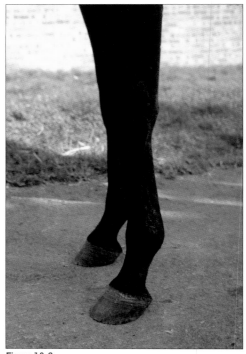

Figure 10.2:
Mismatched Feet; the left fore (nearest) has a broken-back hoof-pastern axis (HPA) and collapsed heels, the right fore has an upright HPA and foot.

Figure 10.3:
An eggbar shoe offers caudal support and protection to the heels. Although there are excellent aluminium shoes available they still require adapting; the bar has been seated out over the frog.

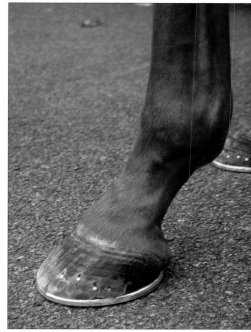

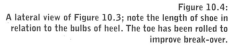

Figure 10.4:
A lateral view of Figure 10.3; note the length of shoe in relation to the bulbs of heel. The toe has been rolled to improve break-over.

knees) and fetlock varus (bowed inward from the fetlock) (Fig 3.21). These may be treated successfully at stud. Some ALDs are ignored or do not respond to treatment. If they are of a less severe nature then they will enter training. ALDs effect both the flight of the limb and weightbearing. The foot always deforms due to uneven weight and becomes unbalanced. Racehorses with deformities caused by ALD are likely to suffer both interference problems and conditions arising from uneven stress to the limbs and feet.

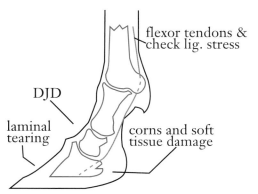

Figure 10.5: Long-toe/under-run heel; causes stress to various parts of the foot and leg.

Lesions to the foot

Injuries and infections can cause a permanent imbalance of the foot. If the foot has been non-weightbearing for a length of time, due to a limb injury or severe infection, it may become more upright and contracted (Figs 8.7 and 8.8). This presents a different shape to the club foot, the solar surface being more oval in shape (the club foot is more pointed and hind shaped). Another reason given for the upright foot is that there is an imbalance in the limb lengths, and the upright foot develops as a result of a shorter limb.

Injuries to the coronary band can destroy the papillae that generate the horn tubules. The hoof wall in that area will then grow horn of a weak and disorganised nature. This type of horn defect is called false quarter. The most common cause of these injuries is from wire cuts. In the UK and US it is still usual to raise Thoroughbreds in post and rail paddocks. In some cases of false quarter the hoof can become distorted due to lack of strength and growth in that area. In severe cases the false quarter may overlap and become a source of infection.

Loss of hoof wall can create an imbalance if the farrier fails to fit the shoe to where the outline of the foot should be. The hoof wall may be lost with a shoe pulled off. This usually occurs where the wall is already stressed, eg a toe-in, fetlock varus deformity which will cause excessive weight to be carried on the lateral hoof wall, making it more difficult to nail and more likely to suffer from risen clenches and deteriorating wall quality. If the shoe is lost and the wall damaged the farrier will be tempted to rasp back to good horn and fit the shoe tighter. In doing so, the shoe will offer even less support to the lateral hoof wall and the imbalance will become persistent.

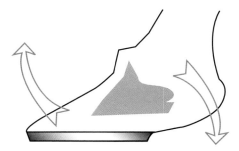

Figure 10.6: Note the point of break-over and posterior point of caudal support in relation to the centre of the coffin joint.

Shoeing

It is possible for a farrier to create an imbalance where one did not previously exist. For instance, if a farrier continually over-dresses one side of the foot an imbalance occurs. Many under-run heels are created, not from over-lowering, but because the farrier believes he is preserving the heel by not

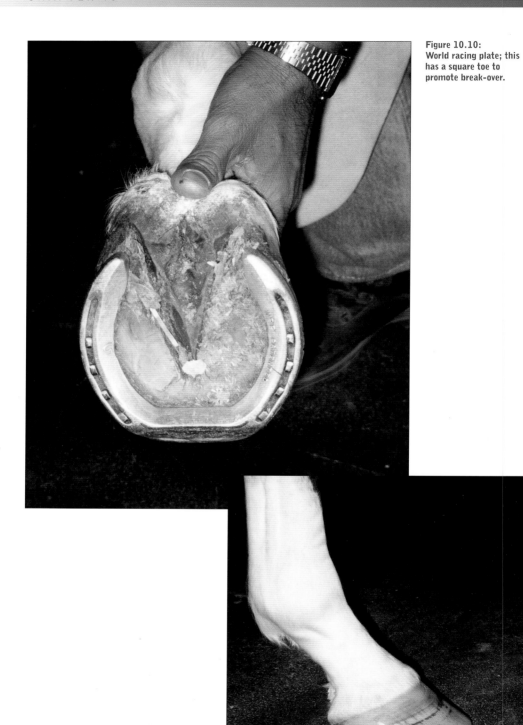

Figure 10.10:
World racing plate; this has a square toe to promote break-over.

Figure 10.11:
A lateral view of Figure 10.10 showing the break-over point back from the toe.

rasping it. Incidentally, club feet cannot be created by a farrier not rasping the heels – it has been tried.

For the most part, farriers do not create imbalances, they just fail to address the minor distortions caused by slight limb deformities. Left unchecked, these worsen until there exists a greater distortion that all can see as an unbalanced foot.

dorsal hoof wall. Vertical projections on the lateral heel, whether they are called mud calks, calkins or stickers, destroy the medio-lateral balance of the foot. When a horse is shod with a three-quarter shoe (fortunately these are now going out of fashion) the foot is distorted out of balance (Fig 10.1).

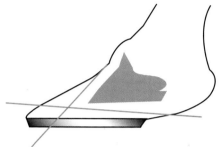

Figure 10.7: The foot is trimmed along the lines indicated.

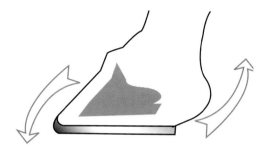

Figure 10.8: The new hoof shape and shoe fit improves the biomechanics.

Purely by placing a normal shoe onto the balanced foot of a horse, we unbalance it. The break-over point is moved away from the tip of the coffin bone and thereafter the relationship of the heel as the caudal point of support is altered. In a natural environment, the horse wears the foot evenly while rolling the toe. When we, as farriers, attach a shoe, the wear pattern of the foot is reversed. The toe is then protected and allowed to grow long, the heels wear by expansion against the shoe. The hoof grows down at the angle of the horn tubules, not vertically. The shoe is therefore moved forward in relation to the limb from the day that it is fitted. The perfect shoeing is out of balance within a few weeks, if not days.

The types of shoes and plates used in racing, by their design, will often unbalance the foot significantly (Figs 9.3-9.13). Light narrow shoes with pencil heels offer little support or stability to the hoof capsule. Where toe-grabs are allowed the HPA is broken-back and there is increased tension at the laminae of the

Types of imbalance

Club foot

The club foot is a chronic condition which cannot be cured. Over trimming of the heels of the club foot may throw excessive strain upon the flexor tendons and the sub-carpal check ligament and result in lameness. That is not to say that the farrier should not attempt to improve the shape. Some judicious trimming of the heels and promotion of toe is advised. Where the dorsal hoof wall is concave it should be dressed back to straight. One balancing guideline is to trim the foot so that the sum of its dimensions match the sum of the dimensions of the normal foot on the other side, eg if the normal foot is 12.5 cm x 12.5 cm and the club foot has a width of 11.5 cm, trim the heels until the length is 13.5 cm. The farrier should try to create a straight hoof-pastern axis (HPA) (Fig 1.9) There should not be an attempt to match the feet in shape, it is difficult enough to nail securely to a club foot without trying to shoe cosmetically. Because of

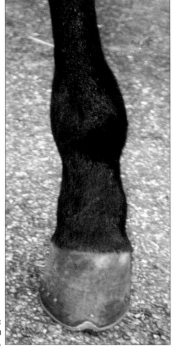

Figure 10.14:
A fetlock varus deformity on the hind leg of a 2-year-old.

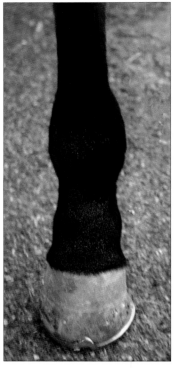

Figure 10.15:
The fetlock varus deformity (Fig 10.14) is shod with a lateral extension to improve leg and foot alignment.

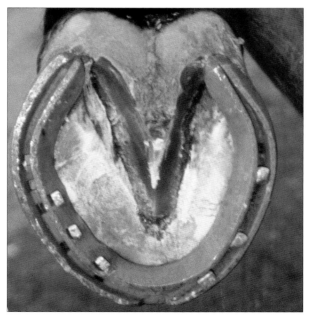

Figure 10.16:
The nail-holes in the lateral extension shoe (Fig 10.15) have been placed on the inside web of a ready-made shoe to allow the wider fitting of the outside branch.

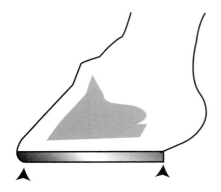

Figure 10.9: Compare the new points of break-over and caudal support with Figure 10.6.

the internal stress to the laminal attachment the club foot is prone to poor quality hoof wall which can cause difficulties when nailing. Seedy toe is often present, compounding any shoeing difficulties. If necessary a hoof repair composite can be used to fill a seedy area.

Upright foot

An upright foot needs alternative shoeing considerations to the club foot. The opposing limb to the upright foot is usually wider and flatter (Fig 10.2), perhaps due to extra weight taken on it when the upright foot was formed by a non-weightbearing injury. The flat foot may continue to carry more weight than the upright foot. If there is a concerted attempt to bring the height of the upright foot down, additional strain may be thrown on the flat foot. It is usual to support the caudal area of the flat foot by shoeing with length or possibly an eggbar shoe (Aluminium; KB Eggbars, GE, California, USA; Steel; Baker Eggbars, Baker Horseshes, Stourbridge, UK) (Figs 10.3 and 10.4).

Long-toe/under-run heels

The Thoroughbred with long-toes and under-run heels is probably the most common imbalance in racing. The Thoroughbred's conformation, notoriously weak horn and type of work lead to a sizeable proportion acquiring this condition. However, no foal was ever born with under-run heels and a concave dorsal wall. The long-toed foot with under-run heels is not inevitable. This hoof shape causes excess stress and strain throughout the foot and limb (Figs 10.5 and 10.6) resulting in:

1. Corns and soft tissue damage in the caudal area of the foot occur due to folding over of the horn tubules at the heel, a concentration of pressure due to a shortening of the heels horizontal length, and the alteration in gait increasing heel landing.

2. Tearing of the distal dorsal laminae (mechanical laminitis) results from the extra stress required to break over.

3. Possible DJD is caused by the broken-back HPA forcing the anterior margins of the coffin and pastern joints together.

4. Excessive tension on the flexor tendons increases the likelihood of strains and damage to the check ligament.

Rebalancing consists of dressing back the toe so that the dorsal wall is aligned. Radical backing up the toe removes the outer hard layer of protective horn. The horn inside is softer and lacking strength and, therefore, should be treated with a hoof hardening agent (Keratex Hoof Hardner, EPC, Somerset, UK). The heels should be dressed back to straight horn, this moves the whole base of support of the foot back towards the leg (Fig. 10.7). Shoeing with caudal support either with long heels or with an eggbar shoe (Figs 10.3 and 10.4). has a beneficial biomechanical effect by moving the foot further under the leg. A rolled or rocker toe will move the point of break-over further back reducing stress to the tendons and laminae (Figs 10.8 and 10.9). Bringing back the point of break-over is achieved by using the so called 4-point method and a plate with a square toe (Figs 10.10 and 10.11).

Bull-nosed foot (Fig 10.12)

The bull-nosed foot is more often seen in the hind limbs. It has a broken-back HPA. The

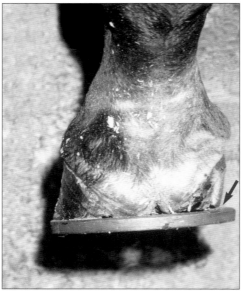

Figure 10.18:
Floating the heel; a section of the hoof wall from medial quarter to heel is removed (see arrow). When the shoe is then attached there is a clear gap between wall and shoe. Usually a bar shoe is used for added stability and to ensure that the medio-lateral balance is maintained at 90°.

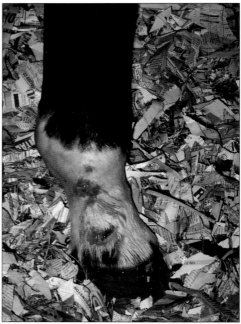

Figure 10.19:
The pastern of the horse shown in **Figure 10.1** showing severe interference lesions caused by scalping.

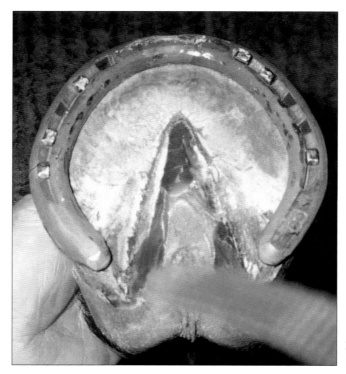

Figure 10.20:
A shoe made from half-round section fitted to a front foot to improve break-over and reduce the injurious sharp edge.

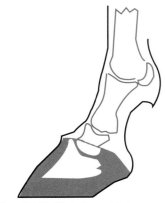

Figure 10.12: The bull-nosed foot.

pastern looks as if it is coming out of the back of the foot. It is always a sign that there is an excess of foot, especially toe. The foot can usually be rebalanced to a straight HPA.

Offset knee

Horses in training with offset knees are prone to a multitude of lesions. They usually have varal fetlocks and suffer from medial splints. The toes flare medially and turn in (Fig 10.13).

Figure 10.13: The offset knee; causing fetlock varus and medial flaring.

When they move they paddle, break-over on the outside toe and land on the outside of their foot. They may acquire a quarter crack on the lateral quarter to heel. When balancing the farrier should be careful to eyeline the leg and/or use a T square. Many textbooks state that a foot with a toe-in imbalance should be lowered on the medial side. In the mature (2-year-old and older) this is not the author's experience. The majority are high laterally. Once levelled to the long axis of the cannon, any medial flare can be dressed back. The shoe can be fitted wide on the lateral side to offer support (Fig 1.13).

Fetlock varus

Fetlock varus is frequently seen in the front limbs with offset knees and is covered above. It is quite common in the hind limbs (Figs 10.14–10.16) where the foot is seen to be more upright on the lateral aspect and flared to the medial aspect. Eyelining shows the foot to be higher medially. This conformation leads to brushing and breaking up of the lateral wall. Rebalancing is aimed at trimming the solar surface of the foot to 90° of the long axis of the cannon and bringing the medial flare in. A shoe should be fitted wide enough to make the shoe symmetrical around the centre axis of the foot (a line through the centre of the frog).

Carpal valgus

The carpal valgus or knock-kneed horse is quite common in training in less severe forms. The horse is usually base-wide and toes out. The feet have probably begun to deform as a foal, to accommodate the medial weightbearing. The action is breaking-over at the medial toe, dishing, landing on the lateral side and impacting hard on the medial side. The foot will be upright or under-run medially and flaring laterally. Eyelining will show that the lateral side of the foot is higher and that the bulb of heel is shunted up. Where the medial bulb of heel is shunted up the distortion is sometimes described as 'sheared'. This is often an inaccurate description unless the 2 bulbs are

Figure 10.21:
An anterior view of Figure 10.20 showing how the half-round shoe is fitted 'tight' under the foot.

Figure 10.23:
A rim rebuild using composite; on very flat feet with poor quality walls repairing the foot in this way gives concavity and moves the nails away from the sensitive structures, making the horse less 'footy'.

moving independently. The older term 'wry foot' may be better. Due to its action, this type of imbalanced foot is very likely to give rise to medial corns and medial quarter cracks. In cases of severe distortion of the hoof capsule medial wing, fractures of the distal phalanx can occur.

Figure 10.17: Floating the heel in a case of severe heel shunting.

To balance the foot, the lateral wall should be flare dressed and the solar surface trimmed to 90° of the long axis of the cannon. The shoe should be fitted symmetrically to the foot. This may mean fitting wide on the medial branch. There is an obvious danger of both shoe loss and/or interference with the other limb. In these cases common sense must be used and adjustments to the individual made.

It is imperative that there is no attempt to equalise the length of the heels. The only method that should be used to bring down the shunted heel is to 'float' the heel (Fig 10.17). To do this, balance as described and fit a bar shoe. Immediately prior to nailing, a section of the hoof wall from medial quarter to heel is removed. When the shoe is attached there is a clear gap between wall and shoe (Fig 10.18). The hoof wall, bulb and coronary band will drop distally within days. This procedure should be used cautiously. Extremely distorted feet will be improved by this method but complete symmetry does not seem to be obtainable.

Interference

'Interference' is the term used to describe when the shoe or foot of one limb strikes another (Fig 10.19). The causes of interference include: poor conformation, exercise surface, rider, immaturity (ie yearling/2-year-old), unfit horse, imbalance caused by poor shoeing, imbalance caused by overgrowth.

Types of interference include:

- Brushing; when the shoe strikes the opposing limb below the fetlock.
- Speedy cutting; when one foot strikes the opposing limb above the fetlock.
- Scalping; when the outer toe of a front foot strikes the hind leg of the same side.
- Cross firing; where the horse interferes diagonally (is more likely seen in trotters).
- Over-reach; where the hind shoe strikes the bulb or shoe of the front.
- Forging; where the hind toe strikes the inside webb of the front toe.

A farrier is limited in overcoming interference by the nature of the causative factor. For all types of interference, shoeing consists of balancing the hoof on the principles already

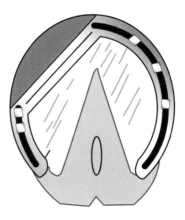

Figure 10.22: The toe preventer shoe.

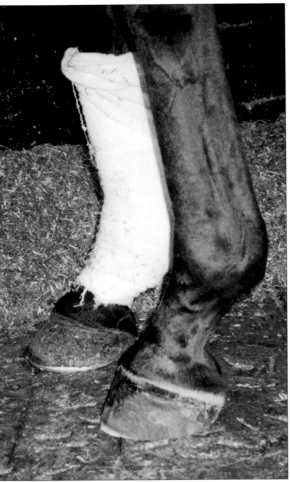

Figure 10.24:
A club foot; the left fore (nearest) had suffered from laminal tearing and a subsequent laminal abscess.

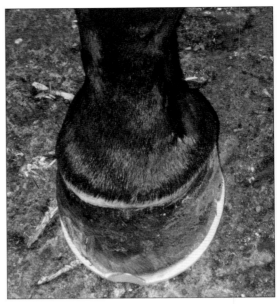

Figure 10.25:
The hoof (Fig 10.24) was rasped back to solid horn, a plate fitted and attached with just the heel nails. Note the area of lesion to the medial (left) quarter.

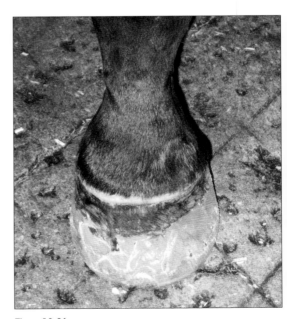

Figure 10.26:
The hoof (Figs 10.24 and 10.25) was repaired with Play-Doh to protect the sensitive tissue from the composite.

stated and removing the portion of the shoe that is causing injury, ie rounding or safeing the edge. Fitting a three-quarter (Fig 10.1) shoe rarely stops a horse brushing and will unbalance the foot. Half-round shoes provide a smooth safe edge and have the added advantage of allowing the foot to freely break over in any direction (Fig 10.20 and 10.21).

In severe cases of brushing, speedy cutting and scalping a toe preventer shoe (Fig 10.22) can be tried. This is very effective in removing the area of the shoe causing the injury. It does have the disadvantage that it may unbalance the foot. The shoe is fitted with the safed portion on the medial branch for brushing and speedy cutting and on the lateral branch for scalping. Fitting the lateral branch of the hind foot wider (lateral extension) and longer, even with a trailer will throw the foot wider. This proves successful in some cases. Shoeing the front foot with a rolled or rocker toe can be effective in promoting a faster break-over in front and therefore moving the limb out of the way of the hind foot.

Many authorities state that slowing the action of the hind foot in cases of forging and over-reaching is achieved by lowering the heels of the hind foot or thinning the heels of the shoe. In the authors opinion this is incorrect in theory and detrimental in practice. Ensuring that all 4 feet are balanced, the front shoes fitted with rolled or rockered toes and the hind toes safed off is the best guard against over-reach. Racehorses usually only forge when trotting, as yearlings or young 2-year-olds, on a deep surface.

Figure 10.27:
A heartbar shoe fitted to a horse in training which had suffered damage to the white line.

CHAPTER 11

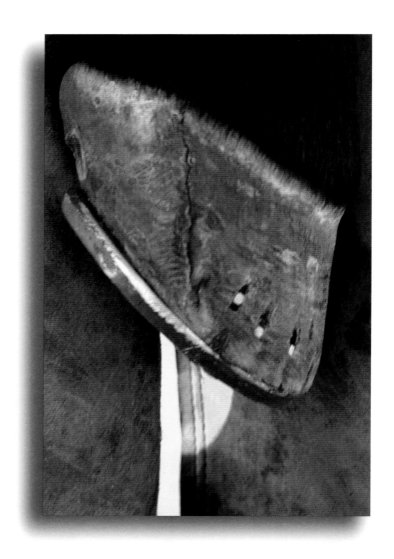

FARRIERY – FOAL TO RACEHORSE

The treatment of hoof cracks and hoof wall lesions

Hoof cracks can occur at various sites and to varying depths in the horn of the hoof capsule. They can be the result of poor hoof horn quality and of foot/limb imbalances. Recognition of the cause of the cracks is fundamental. Successful treatment may be complicated by involvement of the sensitive laminae and infection. Farriery treatment for hoof cracks involves correcting, where possible, foot/limb imbalances, shoeing to reduce hoof capsule movement and immobilising the hoof wall in the proximity of the crack.

11.7 and 12.4). On the solar aspect of the foot, cracks are usually either transversing the bar or sole, radiating from the apex of the frog (Fig 11.8).

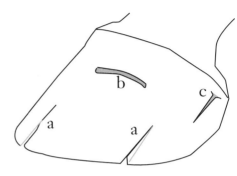

Figure 11.1: Hoof cracks are referred to by their nature: a) grass crack; b) horizontal crack; and c) sand crack.

The nature of hoof cracks

Hoof cracks are referred to by type, location and depth (Figs 11.1, 11.2 and 11.3).

Type (Fig 11.1)

A sand crack originates at the coronary band, continues distally, running parallel to the horn tubules, either completely or partially towards the distal border. They can be thought of as a fracture to the hoof wall (Fig 11.4).

A grass crack originates at the distal border, running parallel to the horn tubules proximally, either completely or partially, towards the coronary band. They can be thought of as a split in the hoof wall (Fig 11.5).

A horizontal crack is parallel with the coronary band. It initiates at the coronary band and grows out with the hoof (Fig 11.6).

Location (Fig 11.2)

Most hoof cracks appear in the wall where they are described as toe, quarter or heel (Figs 11.5,

Depth (Fig 11.3)

Superficial cracks only penetrate the outer insensitive horn. If neglected these can become deep.

Deep cracks penetrate to the sensitive tissue (Fig 11.9). These often bleed during exercise and may become infected. Lameness may result from movement of the edges of the cracked horn causing pinching of the sensitive laminae or sepsis.

Causes of hoof cracks

Grass cracks are caused by poor hoof quality, overgrowth, the environment and seedy toe. Some hooves are of a shelly consistency and are easily split. They are weakened in extreme wet or dry conditions. If allowed to become overgrown the wall is no longer supported by the laminal bond at the white line and increased leverage tears the distal wall open. Hooves can be split by the use of over large nails and nailing too fine into the hoof wall (Figs 11.10 and 11.11).

CHAPTER 11

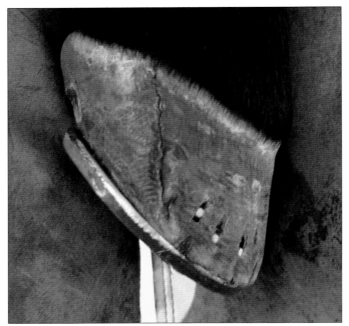

Figure 11.4:
A complete sand crack originating at the coronary band and running to the distal border.

Figure 11.5:
A partial grass crack at the quarter originating at the distal border of the hoof wall.

Figure 11.6:
A horizontal crack, originating at the coronary band, is usually caused by a laminal abscess exiting proximally and disrupting horn production. Occasionally they are caused by trauma.

Sand cracks occur because of uneven stress to the hoof capsule arising from foot/limb imbalance. Contributory factors include those given for grass cracks as well as foot balance, shoe fit and type and the surface and speed at exercise. Although coronary band treads were often given as a cause of sand cracks in working horses in the past, direct trauma to the coronary band is not likely to cause sand cracks in athletic horses.

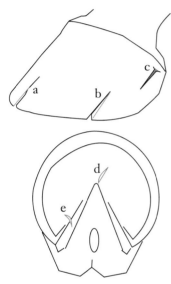

Figure 11.2: Hoof cracks are referred to by location: a) toe crack; b) quarter crack; c) heel crack; d) solar crack; and e) bar crack.

Toe sand cracks are mostly seen where there has been a persistent anterior-posterior hoof/limb imbalance (Fig 11.12). Over long-toes with the break-over anterior to its most efficient point creates an increased lever arm effect. The pull of the deep digital flexor tendon is resisted by the distal anterior hoof wall causing laminal tearing dorsally. The consequence is a concave dorsal hoof wall, a flattened or even convex sole, and a deep sand crack central to the toe (Fig 11.3). Occasionally 2 cracks occur equidistant from the centre of the toe (Fig 11.13).

Without treatment, superficial cracks may become deep and increase the chance of infection. The coronary band becomes damaged and the hoof wall pulls into the crack. The hoof wall is then not a continuous curve and, when viewed from above, dips inwards. Once this occurs the condition is usually chronic.

Quarter cracks and heel cracks are the result of medio-lateral (M-L) imbalances of the hoof and limb. The primary cause is poor limb conformation. This leads to uneven foot fall and therefore stress through the hoof capsule, a consequent distortion leading to a wry foot and ultimately hoof wall fracture. Secondary contributing factors are: 1) the failure of the farrier to recognise the M-L distortion and reduce it; 2) the hoof quality; and 3) firm going and some synthetic surfaces which may increase stress because the horse moves faster on them.

Horizontal cracks originate either from trauma to the coronary band causing temporary cessation of healthy horn growth or where a laminal wall abscess has broken out at the coronary band. They are not a problem unless they interfere with nailing or cause a partial avulsion of the wall. Further infection within them is uncommon.

Bar cracks are frequently seen in cases of low grade chronic laminitis. They are also seen in the long-toe/low-heel syndrome. The fissures are deep and occasionally overlie sub-solar abscesses under-running the heel and frog. In both types of foot this crack may be caused from pain forcing excessive weightbearing on the heels.

Solar cracks radiate from the apex of the frog. They are usually superficial, but can be deep and occasionally infected. They are most often seen in the chronic laminitic and the young horse with an upright or slightly clubby foot. The cause would appear to be the direct

CHAPTER 11

Figure 11.7:
A toe crack.

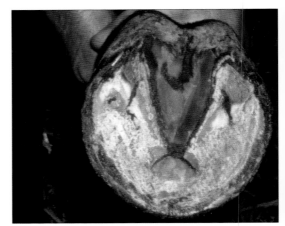

Figure 11.8:
Solar cracks radiate out from the apex of the frog. Bar cracks transverse the bars usually midway.

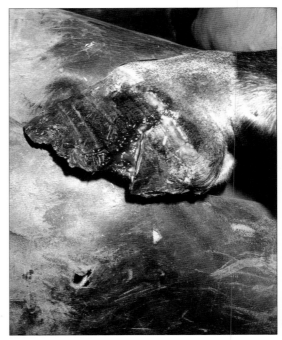

Figure 11.9:
Deep cracks may cause lameness due to infection or pinching of the sensitive tissue.

pressure from a rotating distal phalanx. In the young horse they are temporary whereas, in the laminitic foot they can be persistent.

False quarter is a term used to describe the abnormal hoof wall growth after a severe coronary band lesion (Fig 11.14). The papillae that produce the horn tubules are damaged and the horn grows in a weakened and disorganised manner. The width of a false quarter varies depending upon the size of the original lesion. There is often a sand crack down the centre of a false quarter which may be superficial or deep.

Treatment of hoof cracks

Deep and superficial grass cracks only extending partially up the wall can mostly be trimmed out in routine foot dressing. Flares must be removed and the hoof wall brought back into straight or slightly convex line. Where cracks extend more than halfway up the wall, rasping will further weaken the wall and increase evaporation, leading to a more brittle hoof. Superficial cracks do not penetrate directly into the horn, but run obliquely under the horn tubules. The curved end of a farriers knife will remove the flap without damaging surrounding horn. Scoring the hoof wall above cracks is not successful and tends to weaken the hoof wall (Figs 11.15-11.18). Seedy toe may underlie deep grass cracks and the cavity must be thoroughly debrided and where possible, left open (Figs 11.19 and 11.20).

In severe cases of multiple grass crack, it may be necessary to apply a shoe in order to stabilise the foot. Neither nails nor clips should enter or interfere with cracks. Deep grass cracks should be treated as above with care taken not to damage the sensitive tissue. It may be beneficial in the case of deep grass cracks, especially when at the toe, to fit a shoe with clips at the widest part of the foot. This will reduce over expansion and movement of the hoof. Although radical foot dressing must be carried out to remove flares and correct foot/limb

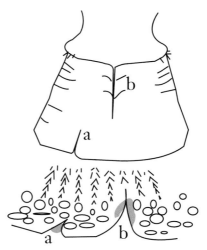

Figure 11.3: Hoof cracks are referred to by severity: a) superficial crack penetrating only the insensitive horn; and b) deep crack penetrating to the sensitive tissue.

imbalances, this will inevitably remove the outer harder layers of horn leading to a further weakening of the horn structure. In order to prevent this, hoof hardener (Keratex Hoof Hardner, EPC, Somerset, UK) should be used over the whole surface of the wall, especially after trimming or shoeing (Fig 11.21). There are various feed additives on the market which claim to improve hoof quality and growth. However, the author has never been convinced of their efficacy.

Infected cracks must be treated immediately. The infection must be allowed to drain and the area to be ventilated. Healthy attached horn should not be removed as this will further reduce stability and impedes capacity to transfix the crack. Radical removal of the horn and sensitive tissue removes the infection but will also complicate repair. The lesion should be thoroughly flushed out with a topical antibiotic.

Treatment of toe sand cracks

Toe sand cracks are generally associated with poor foot balance. The aim is always to correct the foot/limb imbalance, bring the point of break-over back under the foot, gain a straight

Figure 11.10:
Grass cracks may occur when the hoof wall becomes overlong losing laminal support and increasing leverage. (Photo K. Swan).

Figure 11.11:
Shelly hooves cause the outer hoof wall to flake off and may cause nailing difficulties.

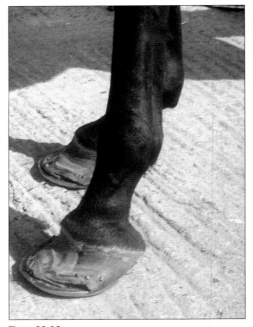

Figure 11.12:
Persistent dorso-palmar imbalance or a laminitic condition is usually the cause of sand cracks at the toe.

hoof-pastern axis (HPA), and remove any concavity in the dorsal wall. In all cases the parallel alignment of the hoof wall and the dorsal surface of the distal phalanx diverge distally. The distal two thirds of the hoof wall can always be dressed back to align with the proximal third of the hoof wall, ie the sensitive structures do not follow the distortions of the hoof capsule. The hoof wall can always be dressed back to a straight or slightly convex line. Lateral radiographs confirm a divergence of the wall and distal phalanx and will give confidence to trim radically. In cases of either upright/club foot and under-run heels foot dressing is aimed at relaxing the hoof capsule and moving the caudal support posterior (Figs 11.22 and 11.23).

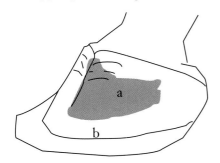

Figure 11.22: Long-toe with a concave dorsal wall and a deep sand crack at the toe: a) position of distal phalanx; and b) correct hoof capsule dimensions.

The shoe type depends on the conformation and severity of the crack. Often a conventional open shoe, suitable for the horses current work, will suffice. The shoe should be changed if not rigid enough to offer stability, eg an aluminium racing plate. The heel type and fit should allow the foot to support the horse on the posterior third of the hoof. Where there is a severe dorso-palmar (D-P) imbalance with a broken-back HPA an eggbar is recommended. A heartbar shoe offers maximum immobilisation of the foot while transferring hoof wall weightbearing to the bone column via the frog and may be useful, therefore, in the most severe cases. Inexperienced farriers should not use heartbars without specialist training.

The type of toe is a matter of choice. The critical factor is the point of break-over. This must be moved back from the toe towards the frog as far as can be safely achieved. A rolled or rocker toe fitted correctly best accomplishes this (Figs 11.21, 11.24-11.26).

Clips either side of the crack are ineffective and possibly damaging for the following reasons: 1) the movement is mostly occurring at the coronary band not at the distal border; 2) with a radical dorsal wall trim the crack will be rasped out in the distal third; and 3) the horn left will be soft and weakened. Clips may be fitted at the widest part of the foot to reduce expansion.

Toe cracks only need to be patched when it is felt that the coronary band needs immobilising to allow healing so that attached hoof wall will grow and replace the crack (Fig 11.27). The same method of patching as shown in Chapter 12 can be used. The patch should be placed as near as is safe to the coronary band after the wall has been dressed back into alignment. Once the patch grows down and interferes with the dorsal wall dressing, it must be removed.

Treatment for heel cracks and quarter cracks

Invariably heel and quarter cracks have the same cause and treatment and unless stated will be considered together.

Priority must be given to correcting the M-L imbalance that has created the initial wry foot and sequential crack. It should be understood that the great majority of wry feet have as their root cause a limb deviation which will not be correctable. The continual uneven stress causes the hoof capsule to distort and finally split. A farrier requires a clear understanding of dynamic foot balance in order to reduce the stresses in the hoof wall.

Although each case of quarter crack has to be assessed individually, it is possible to categorise them and to recognise that certain limb

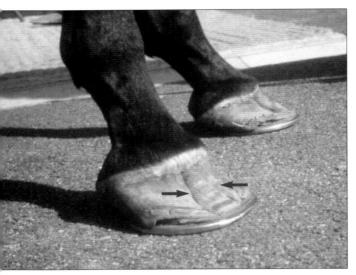

Figure 11.13:
Occasionally toe sand cracks appear equidistant from the centre of the toe.

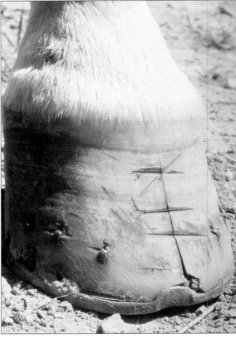

Figure 11.15:
Scoring (rasping) the hoof wall above grass cracks is not a successful procedure. This crack has been repeatedly scored and has continued to grow through each scoring.
(Photo K. Swan).

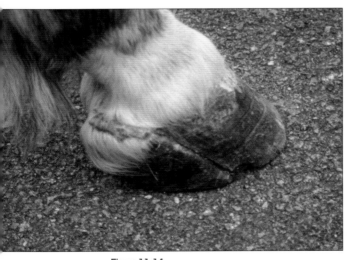

Figure 11.14:
A false quarter is caused by an injury to the coronary band that permanently impairs the production of a normal hoof wall.

THE TREATMENT OF HOOF CRACKS AND HOOF WALL LESIONS

deviations cause predictable distortion of the hoof and therefore a likely site of lesion.

Medial quarter cracks in fore feet are the most common in Thoroughbreds. In an normal to upright hoof they occur further forward than in

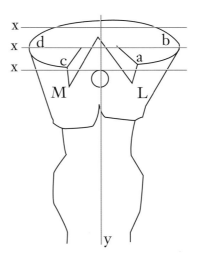

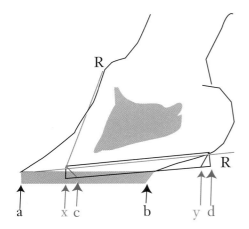

Figure 11.23: Lines R are the correct hoof capsule dimensions, a/b the original points of break-over/caudal support, x/y the corrected break-over/caudal support, c/d further improvement by rolled toe and upright heels.

Figure 11.28: Lines x are at 90° to the cannon, medial heel M is more upright than lateral heel L, the medial heel bulb is shunted proximally, lateral heel quarter b is higher than medial quarter d, heel a is higher than heel c.

a foot with low or under-run heels. When studying the long axis for M-L balance the medial heel and bulb will be shunted proximally. There is often a straightening of the medial side and a flaring of the lateral (Fig 11.28). The medial heel is therefore longer than the lateral heel. The most common farriery mistake is to try to match the heel lengths. Any reduction of the medial heel will worsen the M-L imbalance, causing the foot to contact on the lateral heel and therefore impact harder on the medial heel in the weightbearing phase.

The solar plane of the foot should be balanced at 90° to the long axis of the cannon bone. A bar shoe, appropriately fitted, gives stability to the hoof capsule. The shoe should be fitted symmetrically on the foot, each side equidistant from a line (Fig 11.28 line y) through the frog (Fig 11.30).

Lateral quarter cracks in forefeet are usually found in conjunction with a toe-in conformation. The knees are sometimes offset and the fetlock varal. This conformation leads to an upright lateral wall and a tendency to flare on the medial aspect. Lateral heel cracks are rare and the quarter crack is usually forward though not past the widest part of the foot. These are rare in Thoroughbreds but more common in heavier breeds.

The solar plane of the foot should be balanced at 90° to the long axis of the cannon bone (Fig 11.28). A bar shoe (straightbar or eggbar) with lateral support (ie the distal border of the shoe, caudal to the quarters, wider than the distal border of the hoof capsule above it) should be fitted.

Hind quarter cracks are uncommon but, when they do occur, they are usually on the lateral side in a base narrow horse. Medial quarter cracks are exceptional, generally occurring as a result of cow hocks and/or fetlock valgus. In both cases the foot should be balanced at 90° to the long axis of the hind cannon. For a lateral

CHAPTER 11

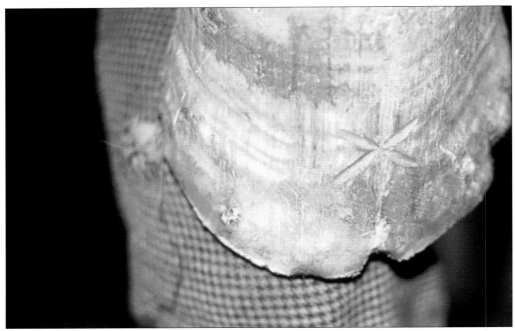

Figure 11.16:
A variation on horizontal scoring is to cut an X in the crack. This is equally unsuccessful. (Photo K. Swan).

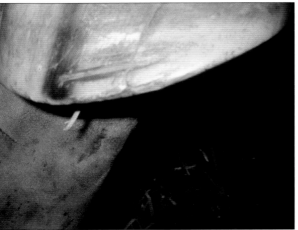

Figure 11.17:
Grass cracks at the quarter are usually the result of flaring of the hoof wall. Scoring the crack has again proved unsuccessful.

Figure 11.18:
The grass crack (Fig 11.17) almost rasps out with correct shaping of the hoof. The crack above the flare is superficial and can be debrided.

quarter crack it is often beneficial to give as much lateral support as possible. With the crack on the medial aspect it is less easy to give medial support. However, some must be given and adequate length of shoe is essential.

Hoof crack repair

Hoof wall crack repair is called patching. There are as many ways to patch a crack as there are farriers to do it. The same principals are basic to all successful methods. The patch must be strong, safe to apply and durable. It must be stressed that without the correct balancing and shoeing, to correct the primary cause, even the most elaborate of patches will not succeed. A traditional method is French nailing. An area of hoof is burned or removed either side of the crack and a nail driven across it. This method entails risk and is not particularly effective (Fig 11.29). The author's choice of patch is by screwing and fibre-glassing or in less severe cases using composite. These methods are described in Chapter 12. Other methods include cross drilling and lacing with umbilical tape and covering with acrylic.

Horizontal crack repair consists of debriding the necrotic horn and filling with a hoof repair composite.

Bar and solar cracks are not repaired and should not be covered by any repair material. Healing occurs with improvement of the causative condition.

False quarter

False quarter is the term used to describe the hoof wall defect that is the permanent result of an injury to the coronary band. The area of affected horn is related to the size and severity of the lesion to the coronary band. The hoof wall grows down in a disorganised and weak manner. There is frequently a crack within the defect which may be deep. Occasionally the wall becomes overlapped.

Treatment of false quarter

In all cases the hoof wall should be kept smooth and continuous with the healthy wall. This can be done at each shoeing with the rasp. It is particularly important where there is an overlap, since these are always prone to infection. False quarter should not be nailed into. If there is a deep crack within the lesion patching may be required.

Partial hoof wall avulsion

Partial hoof wall avulsion is the loss of some hoof wall. It can cause lameness, imbalance and nailing problems. It usually arises from an old over-reach or horizontal crack that has grown down or from shoe loss pulling some hoof wall away.

Repair of partial hoof wall avulsion

Repair may be necessary where shoeing becomes difficult or unsatisfactory due to the hoof loss. Composite repair is the simplest and most effective method. The easy way is to partly nail the shoe onto the prepared foot (not placing nails in the lesion) prior to the repair. This allows the foot to be set down while the composite is curing. Any area adjacent to the repair that needs protecting, eg the nail-holes, should be covered by lanolin (Hydrous Wool Fat BP, J.M. Loveridge plc, Southampton, UK) or filled with Play-Doh (Tonka Corporation) (Figs 11.34–11.37). Where there is a large and severe partial hoof wall avulsion, the foot needs repairing before a shoe can be attached. The composite cannot bear weight until it has cured, therefore it must either be protected or held up during curing.

CHAPTER 11

Figure 11.19:
Seedy toe is often the cause of grass cracks.

Figure 11.20:
The hoof wall and cavity should be debrided where possible to expose the lesion to the air.

Figure 11.21:
Where radical hoof wall rasping is carried out, a hoof hardener may be applied.

THE TREATMENT OF HOOF CRACKS AND HOOF WALL LESIONS

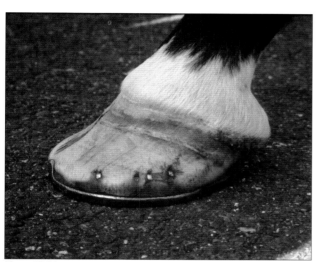

Figure 11.24:
A broodmare with a toe crack caused by a toe flare.

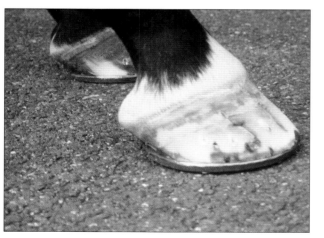

Figure 11.25:
The dorsal wall (Fig 11.24) is dressed back into alignment with the proximal third of the wall.

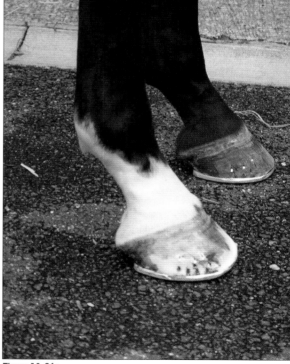

Figure 11.26:
Aluminium rolled toe shoes bring the break-over point back to reduce stress to the dorsal wall.

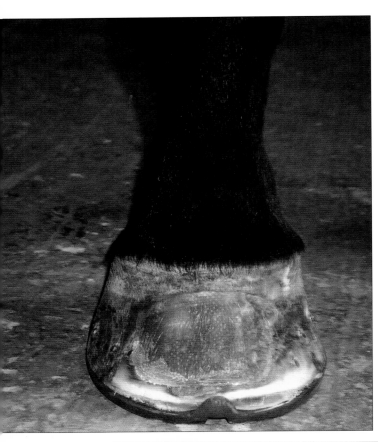

Figure 11.27:
A screw and fibre-glass crack as described in Chapter 12. The patch has remained in place for 2 shoeing periods and sound hoof is now descending.

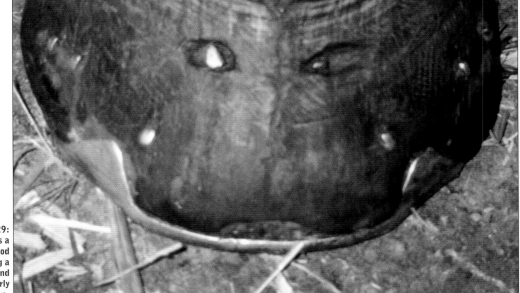

Figure 11.29: 'French nailing' is a traditional method of immobilising a crack. It is risky and not particularly effective.

THE TREATMENT OF HOOF CRACKS AND HOOF WALL LESIONS

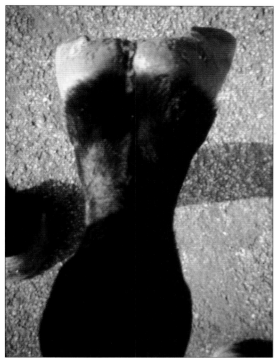

Figure 11.30:
Medio-lateral balance of a hoof with a medial quarter crack (Fig 11.4). The lateral (right) side of the foot is higher. Note that the medial heel bulb is shunted (jammed).

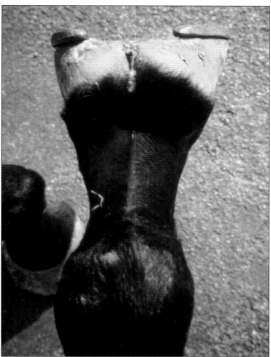

Figure 11.31:
Rebalancing the shoeing (Figs 11.4 and 11.30) is aimed at reducing the high (lateral) right heel and quarter and shoeing with medial (left) support. The base of the foot/shoe is now far closer to 90° to the long axis of the cannon bone.

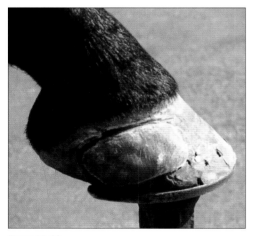

Figure 11.32:
After balancing and shoeing a patch is applied to Figure 11.31.

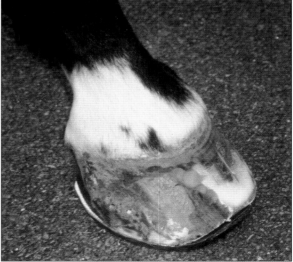

Figure 11.33:
A large area of seedy toe in a broodmare is left open without repair. Unless the horse is unsound this type of lesion is best left exposed to air.

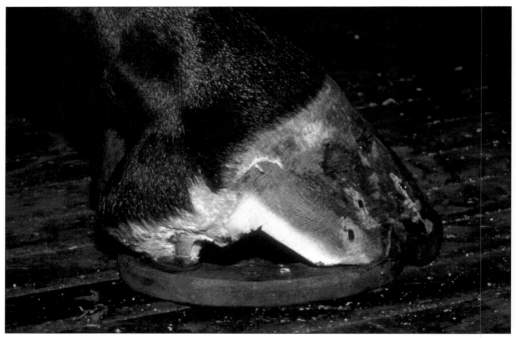

Figure 11.34:
Severe false quarter debrided and shod with a bar shoe.

Figure 11.35:
Lesion depicted in **Figure 11.34** filled with composite.

THE TREATMENT OF HOOF CRACKS AND HOOF WALL LESIONS

Figure 11.36:
A horizontal crack caused through coronary band trauma.

Figure 11.37:
Repair of Figure 11.36 using a composite.

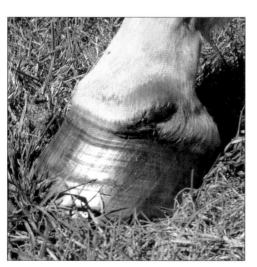

Figure 11.38:
A recent injury to the coronary band of a mare.

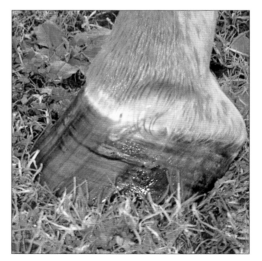

Figure 11.39:
Injury to 11.38 3 months later causing a horizontal crack and a possible permanent defect in the hoof wall.

CHAPTER 12

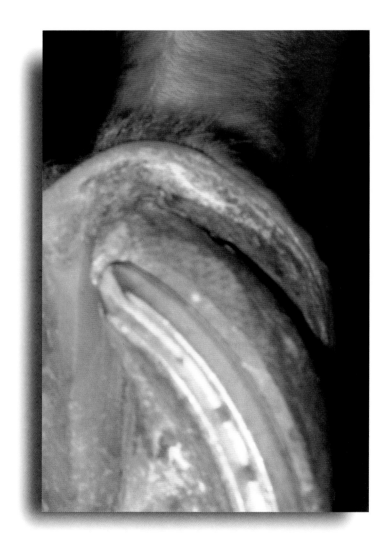

Two methods of patching quarter cracks

Patching is the repair of a hoof crack, usually a quarter crack. There are many techniques used to patch a crack. The same principals are basic to all successful methods. The patch must be strong, safe to apply, and durable. The essence of crack repair is to immobilise the crack while new healthy horn grows proximo-distally. It cannot be over emphasised that the cause of the crack must be addressed to prevent recurrence (Figs 12.1–12.4).

The following are descriptions of 2 types of patches that the author has successfully used on many occasions:

1. The fibre-glass patch

Materials and equipment

Cordless drill
Clippers
Bradawl
Drill bit 7/64"
  (protruding from the chuck by 1/8")
Rasp
Scissors
Screwdriver (with jaws) or screwstarter
Screws (stainless steel self tapping,
  slotted pan head, 1/8")
Syringe (30 ml)
Acetone
Bandage (Vetcon, Vetwrap)
Epoxy Resin 125, Epoxy Hardener 229
  (Pro-Set: Wessex Resins and Adhesives,
  Hampshire, UK; Gougeon Brothers, USA)
Fibre-glass cloth 100 mm × 1000 mm
Lanolin (wool fat)
Mixing tubs
15 panhead, stainless steel slotted screws,
  1/8" × 3 × 16"
Polythene 150 mm × 150 mm
Tongue depressors

Method

Ensure that there is no infection in the crack. If an abscess exists then the foot should be poulticed and treated with an antibiotic. The foot must dry prior to patching. If it is essential to patch the crack immediately, eg the horse is racing the next day and pus is still present, then a drain should be placed behind the crack to allow the application of a topical antibiotic.

A method very similar to this was first described by J. Butler in 1976. The author has found it to be the most effective patch for severe quarter crack (Fig 12.5). The screw pattern described can be varied according to circumstances. A fibre-glass patch can be used successfully on toe sand cracks and, where applied at the toe, the screw pattern should be 4 wide by 3 deep (Fig 11.30).

Trim the hairline above the coronary band, above the crack to a height of 3 cm. Clean the hoof wall surrounding the crack with a rasp and acetone. Do not debride the horn either side of the crack. This will create a wider gap. The screws must be placed as close together as possible for maximum bracing. Drill a pattern of 15 holes, 5 wide by 3 deep. The row of holes either side of the crack must be as close as possible. The proximal row of screws should be as close to the coronary band as is safe.

Fold the fibre-glass cloth into 12 layers, wide enough to cover the holes. Where considered necessary, a drain (a cannular covered in lanolin) can be placed in the crack. Screw the cloth to the wall (Fig 12.6), starting with the distal row of holes. Use the bradawl to find the holes through the cloth. Where it is felt that the proximal and/or caudal rows of screws may penetrate the sensitive laminae

CHAPTER 12

Figures 12.1, 12.2, 12.3
The author has tried many methods of patching over the years. All successful methods of patching must be strong, durable and safe.

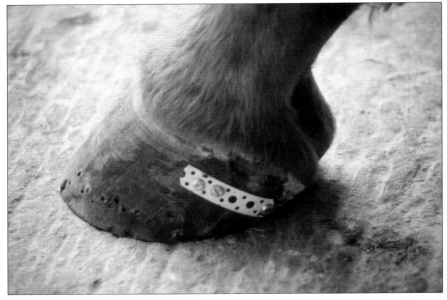

Figure 12.1:
Banding and screwing a patch has tensile strength but allows a shearing action.

Figure 12.3:
Screwing, wiring and covering with a hard acrylic immobilises the crack but the acrylic often cracks.

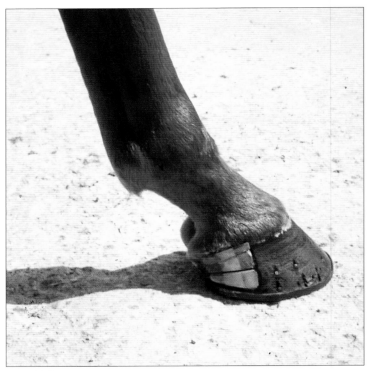

Figure 12.2:
Glueing tabs across allows too much movement.

they should be shortened. This can be achieved by holding them in a pair of pincers and rasping the tips.

Cover the coronary band and pastern with the lanolin for protection. Fold the remaining cloth to 8 layers to cover the screws and ensure that it is to size. This is the top patch. Mix 20 ml of fibre-glass resin to the proportions given in the instructions. Fill the syringe and soak the top patch in the remaining liquid. Use the syringe and tongue depressor to apply the mixed resin to the cloth so that it soaks in (Fig 12.7). Place the top patch over the screws and apply more resin (Fig 12.8). Cover with the polythene and bandage firmly without displacing the top patch (Fig 12.9). After approximately 6 hours remove the bandage and polythene (Fig 12.10). Wipe off the residue of lanolin and tidy the edges of the patch with knife and rasp. If a canular was inserted pull it out at this point. A topical antibiotic can now be injected daily. Tidy the edges of the patch with the rasp. The patch will remain in place for up to 3 months, although it is advisable to remove after 2 months. Avoid nailing into the patch.

NB: If the horse becomes acutely lame on the same foot, remove the patch immediately. There is always a risk of infection when covering any crack. The author's experience is that this occurs in less than 5% of cases.

2. The composite patch

Introduction

This method was first used by the author in 1992 when the first composite material was available. It is an effective patch for all but the most severe quarter cracks. The advantage over the screw and fibre-glass method is speed and less risk. It can also be used in conjunction with correcting medio-lateral imbalance. It may be used as a subsequent treatment to the fibre-glass method while the crack is finally growing out. The author has found it less successful on toe sand cracks.

In the case shown there was both a severe medio-lateral imbalance and some distal hoof wall missing (Fig 12.11).

Materials and equipment

Clippers
Mixing tubs
Rasp
Scissors
Tongue depressors
Acetone
Bandage (Vetcon, Vetwrap)
Composite (Equilox II)
Fibre-glass cloth 100 mm × 600 mm
Lanolin (wool fat)
Milar wrap 200 mm × 200 mm

Method

Debride all necrotic and loose horn and tissue in the area of the crack (Fig 12.12). Trim the hoof wall, sole, bars and lateral sulci to clean horn. Ensure that there is no infection in the crack. Follow the advice given in Method 1 (the fibre-glass patch) where infection is present. Sensitive and/or infected tissue cannot have a composite material laid against it. The foot must dry prior to patching.

Trim the hairline over the coronary band above the crack to a height of 3 cm. Clean the hoof wall, bars and lateral sulci with acetone. Fold the fibre-glass cloth into 8 layers, wide enough to cover the crack and wrap over the sole into the lateral sulci. Check the folded cloth against the foot for position and size.

Cover the coronary band and pastern with the lanolin for protection (Fig 12.13). Fill areas where the composite is not required with Play-Doh or lanolin, eg frog, lateral sulci, nail-holes. Unroll the cloth for impregnating with composite. Use the tongue depressor to apply the mixed composite to both sides of the cloth. Fold the cloth back to 8 layers and squash flat. If necessary, place the drain (cannula covered with lanolin) in the crack. Apply composite to the crack area (wall, sole, bars and lateral sulci)

Figure 12.4:
A quarter crack (arrow) caused by a severe medio-lateral imbalance. The shunted medial heel has led to the crack. For long term success in the treatment of cracks the farrier must address the imbalance causing the crack.

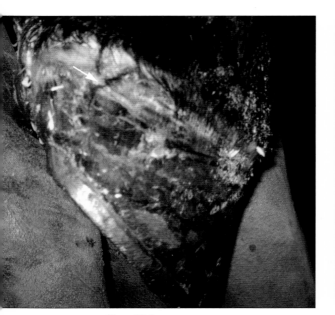

Figure 12.5:
A quarter crack (arrow) that has been treated for infection prior to patching.

Figure 12.6:
A pattern of holes has been drilled; 3 rows of 3 anterior to the crack and 2 rows of 3 posterior. Twelve layers of fibre glass cloth have been screwed to the wall.

TWO METHODS OF PATCHING QUARTER CRACKS

(Fig 12.13). Place the 8 layers of cloth in position ensuring that it wraps around into the lateral sulci. Apply more composite over the cloth and smooth to finish (Fig 12.14). Cover with the Milar wrap and bandage. Smooth firmly to ensure the composite fills all cavities (Fig 12.15).

After approximately 15 minutes remove the bandage and Milar wrap. Wipe off the residue of lanolin. If a canular was inserted, pull it out at this point. A topical antibiotic can be now injected daily. Tidy the edges of the patch with the rasp (Fig 12.16). Nails can be placed in the patch but if possible this should be avoided to maintain strength. The patch will remain in place for up to 3 months. As it grows down, with the wall, it can be rasped as though horn (Fig 12.17).

NB: If the horse becomes acutely lame on the same foot, remove the patch immediately. There is always a risk of infection when covering any crack. The author's experience is that this occurs in less than 5% of cases.

Figure 12.7:
After lanolin has been applied to the hair and coronary band the resin is tamped into the cloth.

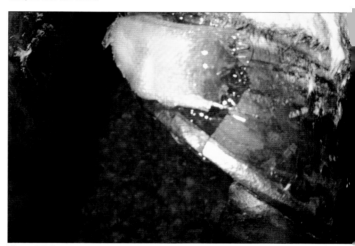

Figure 12.8:
Eight layers of fibre glass cloth are soaked in resin and laid over the top.

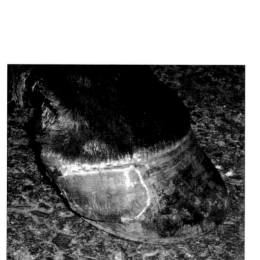

Figure 12.10:
The finished patch is tidied with rasp and knife.

Figure 12.9:
The patch is wrapped in polythene and bandaged for 6 hours.

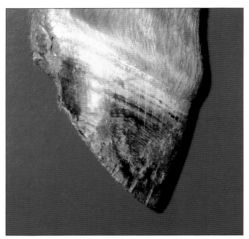

Figure 12.11:
A quarter crack, where there is some loss of hoof wall and a risk of screwing, needs immobilising and rebuilding.

Figure 12.12:
The crack and area to be rebuilt are debrided of loose horn.

Figure 12.13:
The coronary band and areas of sole and frog are protected with lanolin. A layer of composite is applied.

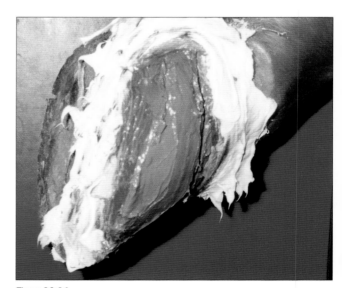

Figure 12.14:
Eight layers of fibre-glass cloth are soaked with composite and applied over the crack and wrapped around the sole and into the lateral sulci.

TWO METHODS OF PATCHING QUARTER CRACKS

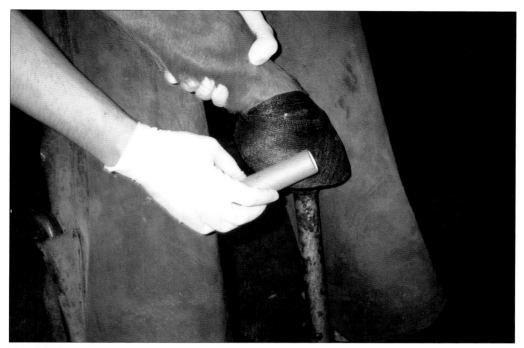

Figure 12.15:
The patch is covered and smoothed into the lesion. The composite will cure in 15 minutes.

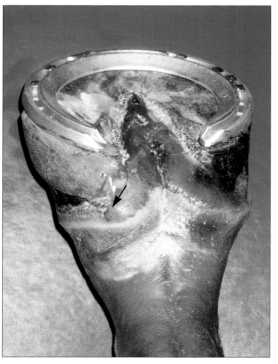

Figure 12.16:
The finished patch is trimmed as required. Note that the crack (arrow) at the coronary band is not covered.

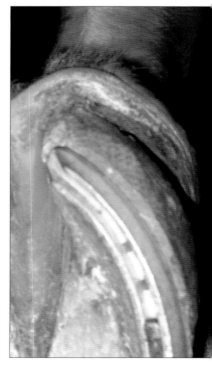

Figure 12.17:
The same patch (Figs 12.11–12.16) 5 weeks later. The new horn above the crack is sound.

FARRIERY – FOAL TO RACEHORSE   129

CHAPTER 13

Shoeing mares and stallions at stud

It is important that mares should have their feet attended to by the farrier every calendar month. Just because they are no longer an athletic animal it does not mean that they do not need healthy, well maintained feet. Their soundness is material to their productivity and they are entitled to their comfort.

Mares usually enter a stud farm wearing shoes in front if they come straight out of training or from the sales. For economic reasons some stud farms stretch the periods between shoeings to 6 or even 8 weeks and this alone is detrimental to the foot (Fig 13.1). There should be an early attempt to remove the shoes and allow the mare to go unshod (Fig 13.2). The alternative is to have a mare dependent on shoes for the rest of her stud life.

There are only 3 reasons to shoe any horse:

1. To protect the foot from excessive wear.
2. To give additional grip.
3. For remedial purposes.

Because, in general, mares are not subject to excessive wear and can grip quite adequately barefoot, the only reason for a mare to wear shoes must be because of some foot or limb problem. When a farrier is asked to shoe a mare at stud he should analyse why the mare needs shoeing and what her problem is. How can shoeing help? Would a different method of shoeing help? Is this a short term condition? What is the goal of shoeing?

Unless it can be seen that shoeing helps then there must be an attempt to have the mare shoeless. When the shoe is removed, the foot should not be over trimmed. It is best done in soft ground conditions, not in the middle of a dry summer or during severe icy conditions. Most mares will show some signs of foot soreness and this is to be expected. If we suddenly went shoeless then we would walk a little more gingerly over rough ground. Give the mare time to adapt (at least one week). If, after this time, the mare is clearly in discomfort then the situation needs to be reassessed.

The conditions of work will affect the level of skill that the farrier is able to demonstrate. Mares should not be trimmed or shod in a paddock with other mares milling around. Neither farrier nor handler is safe. It is quite easy just to take one mare at a time outside the paddock so that she is still in contact with her mates.

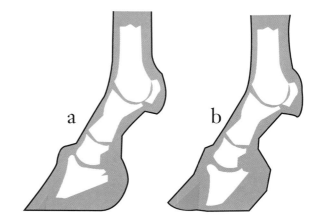

Figure 13.7: a) the low grade chronic laminitic foot; b) the severe chronic laminitic foot.

Trimming

Trimming should be aimed at creating a harmony between foot and limb by removing excessive growth and reshaping uneven growth and hoof capsule distortion (Fig 13.3). Minor lesions should be cleaned out and dressed so as not to worsen. (Chapter 12 covers hoof cracks

CHAPTER 13

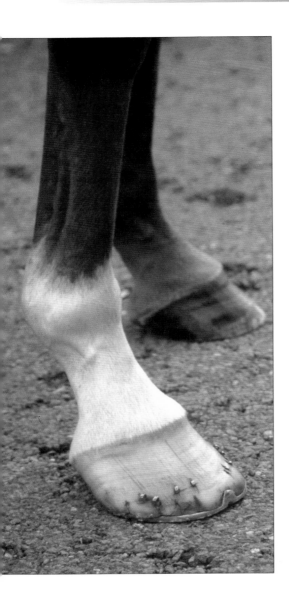

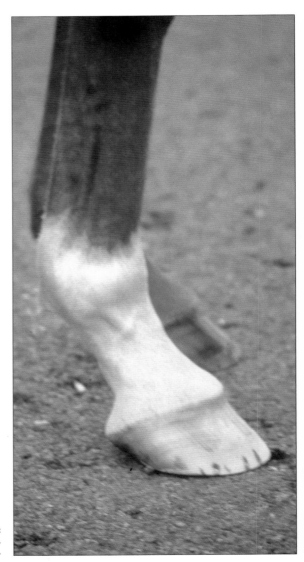

Figure 13.1:
Shoes left on too long are always detrimental to the foot. The shoe is sinking in at the heels and the clenches have risen. The HPA is becoming broken-back.

Figure 13.2:
The shoes of Figure 13.1 have been removed and the feet trimmed. Stud farm mares and stallions should be unshod where possible.

and hoof wall lesions). Frogs should be trimmed. Each side of the frog should be trimmed to remove any flares that trap debris and the central sulcus is trimmed where it hides a cavity. Where present thrush can be treated at this point (Chapter 15).

Trimming the sole is not necessary in a normal foot. The knife should be run round at the white line/sole junction to ascertain the amount of wall to be removed. Sole is only removed where it is below the height of the wall and/or is a false sole. False sole occurs after a long dry spell followed by rain or after a sub-solar abscess (Chapter 15).

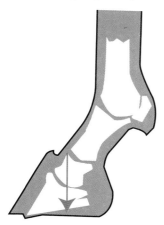

Figure 13.13: A sinker; the coffin bone descends vertically in the hoof capsule.

The wall should be trimmed level to just above the sole and rounded off. Rounding off the trimmed foot stops breaking and splitting (Figs 13.4 and 13.5). Any minor breaks in the wall should not be trimmed out to such an extent as to leave the sole below the hoof wall. The foot should not look like a mushroom when picked up. Minimal rasping of the outer wall should take place. Flares must be removed, but even in severe cases rasping should not exceed more than one third up the wall. There should never be rasp marks left on the foot because these increase the surface area greatly and, therefore, evaporation. They are also a sign of lack of pride in one's craft.

Shoeing

For the previously mentioned reasons, shoeing of broodmares should only occur when necessary. On average only 10% of mares on a stud farm require shoeing. The reasons for shoeing are usually treating hoof wall lesions (Chapter 11) and conditions associated with laminitis and shoeing for sales.

An ideal section of shoe is $3/4" \times 5/16"$ (19 × 8 mm) concave fullered steel. Normal shoeing guidelines should be followed. The foot should be dressed back to suitable proportions and shod symmetrically with full length according to the principles set out in Chapter 1 (Fig 13.6).

Shoeing the low grade chronic laminitic (flatfooted/drop soled)

Most mares that are shod fit into the category of dropped sole or flat footed. These are always related to a level of chronic laminitis (Fig 13.7). Stud farms never describe them as laminitic and usually have an aversion to them being described thus. Nevertheless, it needs to be understood that the reason that any horse has a flat foot or dropped sole is that the distal phalanx is no longer in its correct position in relation to the other structures. This is due to a weakening or loss of the laminal bond between the dorsal hoof wall and the cranial dorsal surface of the distal phalanx.

Signs of the low grade chronic laminitic are:

1. Flaring/concavity of the hoof wall.

2. Cracks, especially dorsal sand cracks.

3. Separation at the toe of the white line.

4. Flat or convex sole.

5. Solar bruising.

6. Prominent bars.

7. Hypertrophy of the frog.

8. Seedy toe.

CHAPTER 13

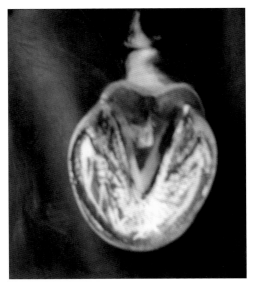

Figure 13.3:
The solar view of a mare's hind foot. The frog has been trimmed and the sole cleaned without over-dressing.

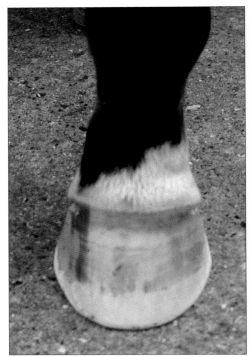

Figure 13.4:
Anterior view of Figure 13.3 after trimming. The distal margin is rounded to prevent breaking and splitting.

Figure 13.5:
Lateral view of Figures 13.3 and 13.4; flare dressing should be kept as low as possible.

Figure 13.6:
An ideal section for mares and stallions is $^3/_4$" x $^5/_{16}$" (19 x 8 mm).

Preparation of the foot for shoeing

The frog should be reduced in size so that with a shoe on there is no direct weightbearing. Where there are prominent bars these should be reduced but not completely removed. The hoof wall should be trimmed to allow for maximum shoe contact. In severe cases of chronic laminitis, this proves difficult with sometimes only the quarter to heels area of hoof wall being sound and healthy (Figs 13.8 and 13.9).

Shoe size and fit

The shoe type and section has already been described. The size of shoe must allow for a full fit allowing for expansion from the quarters back. The heels should be upright and the toe should be rolled or rockered. The shoe must not place direct pressure on the sole and, where necessary, it shoe should be seated out. Where deep seating is needed for a severely dropped sole, a thicker shoe may be required. In order not to overload the limb with weight and the consequent larger shoe nails required, aluminium can be used. Aluminium three-quarter fullered shoes of section 22 x 8 mm or 22 x 10 mm (Kerckhaert Shoes, Stromsholm, Buckinghamshire, UK) (Fig 13.10) can be deep seated cold (Fig 13.11). There is usually a requirement for more nail-holes further back than for a healthy foot. This is because the chronic laminitic foot only has healthy horn in the caudal half.

Shoeing stallions

Stallions are frequently shod because they are exercised on hard surfaces. Usually they are only shod in front although some studs do have their stallions shod all round. In these cases shoeing is conventional. Stallions do, however, suffer from all of the foot conditions that affect mares and where necessary are shod accordingly.

Foot stands

Foot stands allow the farrier to work on the foot pulled forward in comfort. The whole of the outer hoof wall can be viewed at once allowing for a better assessment of shape and a better finish. Very few Thoroughbreds object to their use. They are also a great saver on the farrier's back and knees (Fig 13.12).

Acute laminitis and sinkers

Severe, acute attacks of laminitis are rare in the Thoroughbred at stud. The most common cause would seem to be as a result of metritis in foaling. If the attack leads to sinking then the prognosis is very poor. A sinker is the description of the total loss of laminal bond which allows the coffin bone to drop vertically within the hoof capsule (Fig 13.13). Where there is only a partial loss of laminal bond (at the dorsal hoof wall) rotation of the coffin bone results (Fig 13.14). Where laminitis has lead to sinking then the author has been unable to find any farriery method that works and can claim no success. In cases of rotation of the coffin bone the author has had considerable success using the heartbar and dorsal wall resection method as expounded by B. Chapman. Any farrier who has not had direct tutoring in this method should not attempt it (Figs 13.15-13.17).

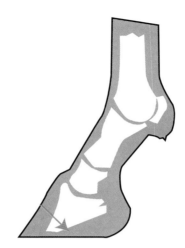

Figure 13.14: Acute laminitis; may lead to the coffin bone rotating within the hoof capsule.

CHAPTER 13

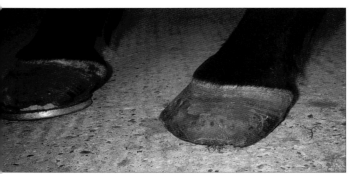

Figure 13.8:
A chronic laminitic mare's foot. The hoof wall at the heels is growing normally. The dorsal wall is detached. The sole is below (dropped) the wall.

Figure 13.9:
Solar view of Figure 13.8 showing sound hoof wall from the quarters to heels. The wall anterior to the quarters is poor to non-existent for supporting a shoe.

Figure 13.10:
An aluminium shoe allows for cover (width) and depth without adding to weight.

SHOEING MARES AND STALLIONS AT STUD

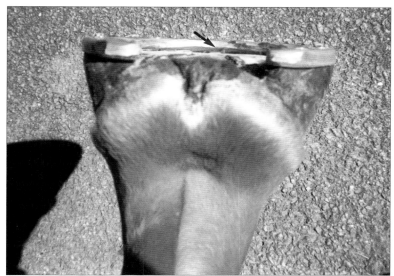

Figure 13.11:
An aluminium shoe can be easily seated out (arrow) for a flat or dropped sole.

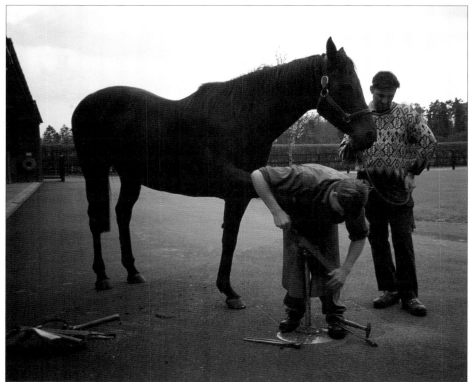

Figure 13.12:
Using a foot stand allows the farrier to see more of the foot - and saves his back.

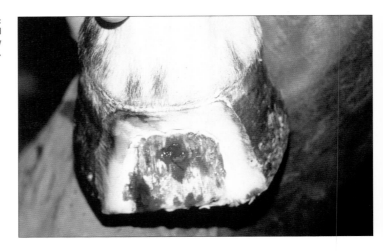

Figure 13.15:
A dorsal wall resection. This should only be used with veterinary involvement.

Figure 13.16:
A lateral radiograph of Figures 13.15 and 13.17.

Figure 13.17:
A heartbar fitted to a hind foot (Fig 13.15) that has had a dorsal wall resection.

SHOEING MARES AND STALLIONS AT STUD

Figure 13.18:
Adjustable glue-on heartbar shoes. Shoeing for laminitis has many variations; most involve bringing back the point of break-over and frog support.

Figure 13.19:
'French bar' shoes for caudal and frog support. The shoe on the right shows where the frog is in relation to the bar.

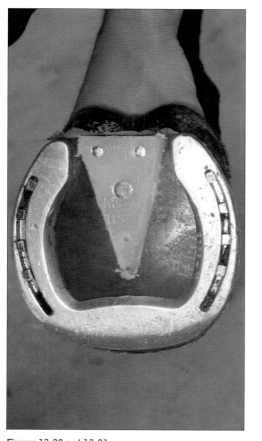

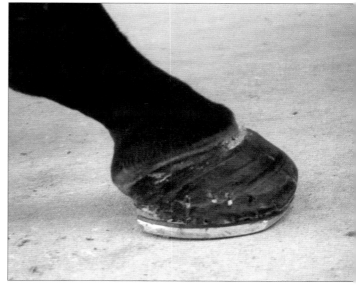

Figures 13.20 and 13.21:
A method advanced by G. Ovnicek for laminitic conditions. The break-over point is radically brought back and a pad with frog support is fitted. Underneath the pad is dental impression material to provide cushioning and spread the load evenly in the caudal half of the foot.

CHAPTER 14

140 FARRIERY – FOAL TO RACEHORSE

The farrier's relationship with the horse, its trainer, owner and veterinary surgeon

Farriery is unusual in that it deals directly with both animals and people. Shoeing horses is hard work. It can be dirty and injurious. It can also be rewarding financially and, perhaps more importantly, in terms of personal satisfaction. Good relationships with the horse, its handler and others involved make life much easier. This chapter considers some of the problems which may sour relationships and how those pitfalls can be avoided or overcome. It is hoped that any advice found in it will, through an improved service to clients, ultimately benefit everyone. The author could not miss the opportunity to also discuss relationships with veterinary surgeons and other farriers.

Professionalism could be said to be an attitude that one shows towards clients and colleagues. This encompasses reliability, presentation and behaviour. How we see ourselves and how others see us matters a great deal in our self esteem. No-one expects their farrier to turn up in a new suit, driving a gleaming car. However, smart clothes and a relatively clean working vehicle go a long way to make a good impression. Our speech and demeanour can either confirm or destroy that impression.

Reliability is perhaps the biggest grievance as far as most clients are concerned. Everyone has at some time been kept waiting. It is a most annoying experience. There is never a time when it is excusable to be late without the client having been forewarned. The farrier's day may never be completely predictable, but it can be made far more so than the chaos that passes for many work schedules. Any young farrier setting out to build up a clientele would be well advised that reliability will gain you most work; good manners, always the cheapest but most valuable commodity, is the next most important and, lastly, good quality work. This may reflect the value that the society we live in places upon good manners, or it may be a sad reflection upon our profession.

Difficult horses

Horses that are difficult to shoe are always likely to place a strain upon relationships. A sympathetic understanding of basic horse behaviour is essential. In nature the horse is not a hunter but rather one of the hunted. His acute senses are tuned to an instinctive reflex system. The horse uses his speed to escape. His only form of aggressive defence is kicking or biting. This means that when you take hold of a leg or foot it is helpful if the horse recognises you as a friend and not an enemy looking for a meal.

The farrier's point of view is usually, and quite rightly, that his service to the trainer should not include 'breaking in'. This is not to say that he does not expect to have some behavioural problems, especially with youngsters. Some trainers and farm managers believe that it is solely the farriers responsibility to shoe unruly horses. They make no effort to improve the situation. In such circumstances a farrier must decide whether it is worth continuing to shoe their horses.

When a farrier shoes a difficult animal they usually work as fast as possible and concentrate on their own safety. Consequently, even the

CHAPTER 14

Figure 14.1:
A shank (or lipchain) is a brass chain that is passed through the headcollar and over the nose.

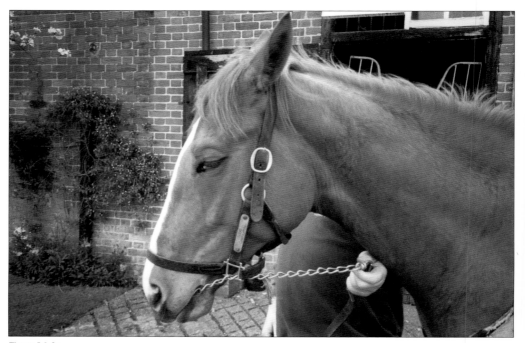

Figure 14.2:
The shank can be used over the upper gum and under the lip or as a bit in the mouth.

most conscientious craftsman will be unable to turn out his best work. It is clearly in the farrier's, the trainer's and naturally the horse's best interests for farriery to be carried out in a calm atmosphere and safe environment.

Restraints

On the race track in the USA a plater would not dream of shoeing a horse that was not held. Unfortunately, in the UK, most trainers are loathe to supply such help. The farrier either employs a holder or ties it up in the stall. A holder should ensure that the horse stands '4-square' and is kept occupied. A shank (brass chain) is useful for this purpose as it can be used as a simple lead, across the nose (Fig 14.1) or across the upper gum under the lip (Fig 14.2).

The twitch works by stimulating the production of endorphins in the brain. These bring on a euphoric state in most cases. The farrier can tell whether the twitch has worked by the look of the horse and usually it becomes quite rigid, needing some force to raise the feet. In this state the horse will usually behave. Nevertheless, the farrier should never be complacent. Horses can suddenly become very agitated by the twitch and strike out.

Chemical sedation is an excellent option today. Acetyl promazine (AcP) (C-Vet Veterinary Products, Lancashire, UK) is often used for taking the edge off difficult horses and young stock. It is usually administered in tablet or paste form. It will not overcome vicious or dangerous horses and, therefore, farriers must never assume that they are safe with a horse simply because it is treated with AcP. Mixtures such as detomidine and butorphanol have by far the greatest effect on even the most unruly horse. These sedatives are administered by intravenous injection, and being controlled drugs have to be held and administered by a veterinary surgeon. The horse enters a trance-like state and remains so for about 40 minutes. Beware that they can still kick and they come round very rapidly.

Young stock

The handling of foals while trimming is a skill in itself. Ideally the farrier has the assistance of 3 experienced pairs of hands, one to hold the mare, a second to hold the foal, and the third to pass tools or to take the tail of the foal if needed. After viewing the foal (Fig 14.3), it should be brought inside a clean loose box with some bedding for safety. The foal is then held alongside the wall. It should not be pushed against the wall as this unbalances and frightens the foal. The mare should face its offspring. This settles the mare and therefore the foal (plus it aims the rear end of the mare away from the farrier, which settles him; Fig 14.4).

Take care when approaching the foal. The majority of farriers develop over the years a natural, slow approach to horses. Unless they trim many foals, they may still move too fast for a youngster. Another frightening thing to a foal is the feeling of being trapped, so when the foot is picked up it should be done so lightly, allowing the foal as much freedom as possible. It is best to trim a foal's foot without placing it between the legs. Holding the pastern with the left hand, most trimming can be achieved by using a sharp drawing knife before finishing with a rasp. This technique is not so easy on the off-side. Dexterity needs to be acquired by practise.

Should the foal 'play up', then the third holder's assistance is called for. He takes hold of the tail as close to the body as is possible. Some holders take a grip halfway down the tail and also distance themselves from the animal. They usually do this out of fear of being kicked. They are a danger to themselves and everyone else. If the holder is not afraid but merely inexperienced then the farrier must educate them.

The shoeing and trimming of yearlings and other immature horses also needs careful attention to safety. Remember that safety is like your money. No one looks after yours better than yourself! The 'head and tail' system of

CHAPTER 14

Figure 14.3:
A foal is assessed by the farrier outside on a clean level surface while the mare is also held.

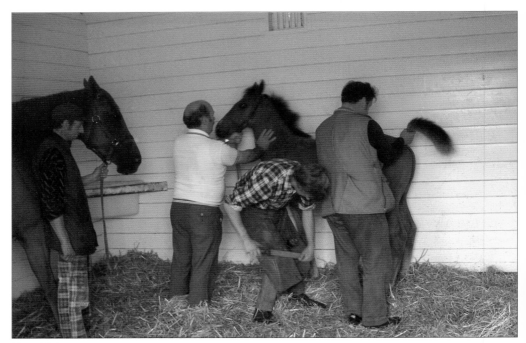

Figure 14.4:
The young foal is held by 2 experienced handlers. There is also a handler for the mare.

holding does not work with larger and older, but still immature horses. Yearlings are like adolescents, they push our good nature to the limit. Remember that however satisfying in the short term, giving him what you think he deserves is only making a rod for your own back.

There is always more than one way to skin a cat. Do not insult your own intelligence by thinking any horse can outwit you. There are a hundred and one alternative methods you can use. Try a bridle, a shank, or a twitch; try giving him something to chew on or do anything that distracts him. Maybe he just wants to see his mates. The list is endless. Only experience enables you to recognise the best technique to try first.

Working with veterinary surgeons can be the most satisfying part of the job. Unless the veterinary surgeon is an exception or you are a very poor farrier you will know far more than him when it comes to horseshoeing. Do not threaten his status by trying to show your superior specialised knowledge. Let him use you as his expert or technician. Remember that while you are working on a case for him he bears the responsibility for your work. It takes a little time to build a good working relationship with a veterinary surgeon but, as a result, you both reap the benefit. Many profitable, prestigious and interesting cases are soon referred once a veterinary surgeon trusts a farrier's ability and professionalism.

The relationship between farriers is frequently very poor. Criticism of one farrier by another, justified or not, in the company of horse trainers, veterinary surgeons and other farriers occurs all too often. We are all trying to earn a living, but none of us needs blow out another's candle to make our own shine brighter.

Deliberate price undercutting will not increase the number of horses to shoe in an area. It might increase a farrier's share, albeit at a lower profit. It is unlikely to attract the best clients. If they go to a farrier because he is the cheapest then they will probably leave him when an even cheaper farrier comes along. It is ironic that our best advertisement is often the price that we charge. Undercutting and bad mouthing colleagues will also alienate a farrier. All of us at some time require help of others. New clients are often gained as a result of recommendation from other established farriers.

Communication with the trainer or stud manager is essential when there is a problem requiring therapeutic or corrective shoeing. The first necessity is that the farrier knows exactly how he/she is going to deal with the problem. If the probable length of time and cost required for treatment is made known, then one potential future source of argument has been removed. Explain clearly the problem and how it can be helped. If the farrier does not know then they should seek advice or if necessary call in someone who can help. Credibility is not lost by acting professionally. Doctors and veterinary surgeons have always sought second opinions.

Education is the key to improvement of relationships within farriery. This involves explaining to the trainer and stud manager how to gain the optimum soundness for their horse by providing the farrier with the best possible conditions in which to work. It also involves convincing the veterinary profession of the achievements possible when there is a true partnership with their farrier. Finally, continuing education for the farrier is essential if the craft is to be lifted in status.

Farriers must be constantly reminded of the advantages to themselves of higher ethical standards. This can be achieved by ensuring that such ideals are promoted throughout a farrier's apprenticeship, and at farriery schools. After apprenticeship, continued association with other farriers leads to co-operation that benefits both the farrier and the horse.

CHAPTER 15

Common conditions of the foot

Thrush

Thrush affects the frog and bulbs of the heel. It is caused by accumulation of debris and infection with anaerobic bacteria in the lateral and central sulci of the frog. Some horses are more prone than others. It is often stated that it is caused by unhygenic conditions. That would not explain why some horses have thrush in just one foot or how, in a stable of possibly 100 racehorses, all kept in the same conditions, only one or 2 are affected.

The farrier's main responsibility is to trim the frog correctly at each shoeing. The frog should be trimmed into a 'V' shape with the sides cut back to new horn. The solar surface need not be trimmed unless it is over-developed or ragged. The central sulci should also be kept clear of ragged horn.

Treatment for thrush consists of trimming as above and removing any necrotic horn (Fig 15.1). This may expose sensitive tissue. The open lesion should then be treated with an oxidiser such as hydrogen peroxide. An alternative is a topical antibiotic, eg sprayed with an antibiotic aerosol (Terramycin Aerosol Spray; Pfizer Ltd, Kent, UK). The lesions are then plugged with cotton wool (Fig 15.2). A final spray ensures that the cotton wool is soaked with the antibiotic (Fig 15.3). This keeps the lesions open and allows in oxygen. It also excludes more mud and faeces. If the process is repeated every 2 days, even the severest cases of thrush can be cured in a few weeks.

Hoof testers

In careful hands, hoof testers are the best way to locate an abscess. A good technique is to start lightly and at the furthest point on the foot from the suspected focal point of infection. The foot is then squeezed, moving methodically toward that point. Once there is any hint of a reaction the testers should be placed on the other side of the suspected area, again as far away as possible. Working back to the suspected focal point of pain will confirm its location.

If hoof testers are applied with force immediately, they can produce such pain that the horse then reacts wherever they are used, giving no chance of obtaining a precise location. It should be remembered that even the soundest foot will react if squeezed hard enough. Common sense should also be used when employing hoof testers. If the horse is only slightly lame, then they may require some force to get a reaction. On the other hand, if a horse shows signs of great pain then if the problem is in the foot only light squeezing will be needed to get a reaction.

If a prick is suspected and there is a sign of an entry point, this can often be investigated with a sharp probe. By moving the probe around, the abscess often bursts.

It must always be remembered that not all lameness' and abscesses can be detected initially and the assumption that 'there's nothing wrong with the foot' should never be made. If there is no reaction then the best statement is 'I cannot find anything in the foot'.

Searching the foot

1. Remove the shoe using nail pullers to extract each nail individually.

2. Clean all areas by trimming a fine layer of horn off the sole bars and white line.

3. Trim the frog and any excess hoof wall.

4. Examine and investigate any obvious lesions.

CHAPTER 15

Figure 15.1:
A severe case of thrush. The sensitive tissue is exposed and there are deep lesions in the lateral and central sulci.

Figure 15.2:
All necrotic tissue should be trimmed off and a topical antibiotic applied. Any cavities should be packed with cotton wool.

Figure 15.3:
After packing with cotton wool the lesions should be treated again. This should be repeated every 2 days.

5. Methodically use hoof testers.

6. Probe and ventilate suspected area of abscess.

Abscesses

Abscesses are an accumulation of pus beneath the outer horn either under the sole (sub-solar) or behind the hoof wall (laminal). Sub-solar abscesses are caused by a nail prick behind the white line or a penetration by a foreign object such as a flint (Fig 15.4). They can also be the result of bruising especially at the seat of corn. The use of fillers over cavities always carries the risk of creating an abscess. Treatment involves drainage and topical antibiotics. If an abscess is suspected, initial treatment is usually with a poultice. This draws the abscess which breaks and allows for ventilation and treatment.

The careful use of hoof testers is the best way to locate an abscess and allow an exact opening above it. Many racehorses lose valuable time in training and miss racing opportunities by aggressive removal of horn and sensitive tissue by the overzealous use of the knife. There is rarely any need to draw blood in finding and treating an abscess. If an abscess has been found there is no point in either applying another poultice or cutting deep into the surrounding sensitive tissue. Another detrimental method of searching for an abscess is cutting a trench along the white line, sometimes even from heel to heel. The white line bonds the hoof wall to the sole, its removal weakens the whole structure of the foot. The farrier will already have to replace a shoe avoiding the abscesses area for some months. The task can be made almost impossible if the foot is damaged in other areas.

Treatment of sub-solar abscess

Once an abscess has been located and drained treatment is aimed at returning the horse to normality as quickly as possible. If it is a horse in training then it is imperative that as few days as possible are lost. Broodmares and stallions need their own form of exercise. In order to protect and reduce the chance of re-infection a protective pad can be fitted under a shoe.

An abscess may be caused by a deep puncture wound. If this has infected the coffin bone septic osteitis may result in a sequestrum (Fig 15.5). In these cases it is essential that the whole of the infected area is debrided and treated (Fig 15.6). A hospital plate shoe (Fig 15.7) offers protection and easy management (Fig 15.8). Once an abscessed area has keratinised over it is safe to fill and cover with a pad.

Pads

A pad is usually made of either leather, metal (thin steel plate or aluminium), or various synthetic plastics. The author's choice for covering an abscess is a hydroplastic pad (TAK Pads; TAK Systems, Massachusetts, USA). This can be moulded to the contour of the foot, thereby excluding infectious matter from the area of the abscess.

The ventilated area of abscess should be clean. It can be flushed with a topical antibiotic, eg Pevodine Iodine (C-Vet Veterinary Products, Lancashire, UK) or Terramycin spray and the area covered with cotton wool soaked in the antibiotic (Figs 15.9–15.11).

The thermal pad is heated in hot water until transparent and laid over the whole of the solar surface and partially across the bulbs of heel. It is firmly moulded into the sulci and the shoe pressed into position. The shoe is nailed on as the pad cools and hardens. Excess pad can be trimmed off and finished with the rasp when the foot is clenched.

Corns

A corn is an area of bruising between the bar and the hoof wall at the heel (the seat of the corn'). Causes of corns include:

1. Short shoeing, ie shoes too small.

2. Shoes left on too long.

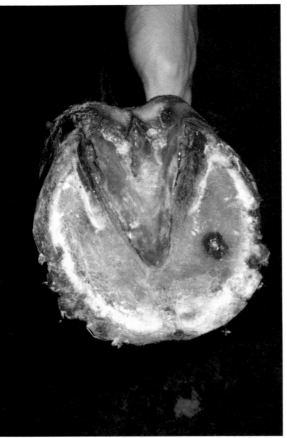

Figure 15.4:
An abscess caused by a penetrating wound.

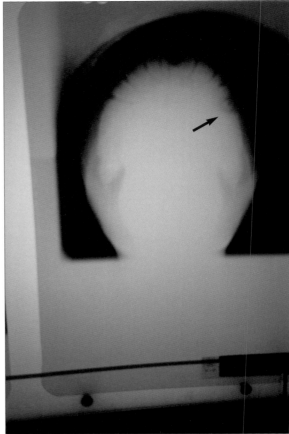

Figure 15.5:
A radiograph of Figure 15.4 shows a sequestrum on the distal margin of the coffin bone.

Figure 15.6:
All necrotic tissue is debrided and the wound thoroughly cleaned.

3. Medio-lateral imbalance causing weight to be carried medially.
4. Dorso-palmar imbalance, ie broken-back HPA and/or under-run heels.
5. Upright HPA.
6. The use of wedges to correct an HPA imbalance.

Corns can be recognised by localised pain and red discolouring of the horn. In severe cases corns can be damp (suppurating) or infected, ie abscessed. Since corns are a symptom of shoeing or conformation, they are treated by attention to the balance and type of shoe. If infected this must be treated prior to reshoeing.

Most corns are not caused by shoes left on too long or short shoeing. Most corns are due to internal forces caused by imbalance, either induced by shoeing or due to the horse's conformation. Nonetheless, good farriery will improve the condition in the majority of cases.

Seedy toe

Seedy toe is the destruction of the inner third of the hoof wall. It may destroy the bond between the sensitive and insensitive laminae. The signs of seedy toe are:

1. A cavity at the white line.
2. Mealy, fetid horn.
3. Grass crack in otherwise healthy hoof (Fig 11.12).
4. Distortion, towards the centre of the foot, of the white line.

Seedy toe does not necessarily occur at the toe and may be found anywhere along the white line, from heel to heel (Fig 11.36). Lameness rarely occurs unless the cavity reaches the coronary band or becomes packed with dirt, etc. It is often a consequence of laminitis. Other causes are less well understood. It may be that seedy toe occurs where the wall has been over-stressed or where damage has allowed bacterial and fungal infections to gain entrance. The cavity is frequently triangular in shape with a deeper fissure penetrating the sensitive laminae in the centre.

Treatment

Exposure to air is considered to be the most effective treatment for seedy toe (Fig 11.22). In many cases this is not entirely practical due to nailing requirements. Where possible, the hoof wall should be removed around the seedy area and all the loose infected tissue debrided. Where the hoof wall is left, the loose horn should be scooped out and the cavity filled with cotton wool soaked in Pevodine.

There are various propriety treatments available on the market that claim to combat seedy toe. The author has not been convinced of their efficacy. Most seedy toe cavities will grow out if treated in the manner described. However, some persist despite dedicated treatment.

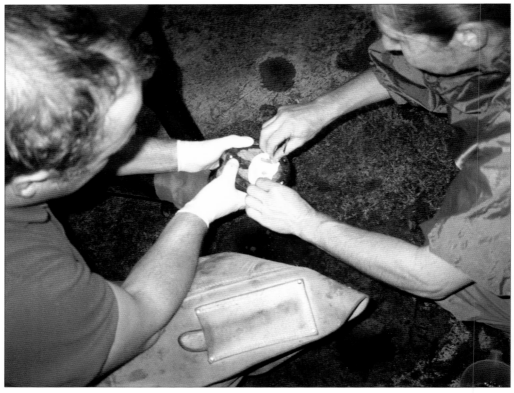

Figure 15.7:
A hospital plate (treatment plate) shoe is fitted and the wound is packed with iodine soaked gauze.

Figure 15.8:
The hospital plate is bolted on and the wound dressed every day.

COMMON CONDITIONS OF THE FOOT

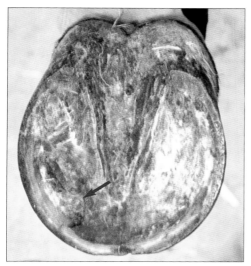

Figure 15.9:
An abscess (medial toe quarter) has been treated leaving a thinly cornified layer susceptible to further penetration and infection.

Figure 15.10:
The area of lesion (Fig 15.9) is cleaned and filled with gauze and antibiotic.

Figure 15.11:
A thermoplastic pad is moulded to the sole to give a good seal and a shoe nailed on.

Figure 15.12:
A severe abscess in this foal's foot infected the coffin bone which is exposed (arrow) and led to it being euthanised.

CHAPTER 16

Forging bar shoes

Given practice and tuition anyone can forge shoes to a reasonable standard. The best tools should be used and these must be maintained. The turning hammer must be suitable for making light shoes (2 lb approximately). The tongs should also be light, but more importantly they must hold the section of barstock securely. Stamps and pritchels must be set up for the appropriate nails and not bulge the section.

The secret to all fire-welding is cleanliness and an even heat (Fig 16.1). The preparation of the 2 surfaces (scarfs) for welding must be thorough. With the advent of chemical fluxes (EZ Weld, Cecil Swan Products, Gloucestershire, UK) a perfect fire-weld can be achieved in gas. Because we all differ slightly in our forging techniques we stretch the shoe by different amounts. Therefore the farrier should learn, by trial and error, how much metal he or she needs. The measurements given below for making bar shoes work for the author, they are not written in stone.

Measuring the foot

The test of any good shoe maker is the ability to make a shoe that is the correct size and fit. The only sure method to do this is to measure the foot accurately using the same method each time and for every type of shoe. The points that are measured to (Fig 16.2) are fixed points. That is why the frog length (F) is taken from the point of buttress, which is close enough to the back of the frog and does not change. Any other point is nebulous and cannot be repeated.

Method of forging a bar shoe (straightbar shoe)

Introduction

The bar shoe is sometimes called the straightbar shoe to differentiate it from other

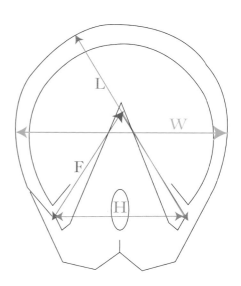

Figure 16.2: Measurement points of the foot: W) the width of the foot; L) the length of the foot; H) the width of the heels; F) the length of the frog.

bar shoes. It is the standard bar shoe. It is the same shape and size as a conventional open shoe with the shoe continuing directly across from heel to heel. Often there is a small frog-plate (widening in the centre of the bar). Sometimes the bar is bent forward to the middle. The bar shoe is used in the treatment of various forms of hoof lesions, eg quarter crack, corns, partial hoof wall avulsion. It can be used for medio-lateral imbalances and in some dorso-palmar imbalances where an eggbar shoe is not appropriate. A barshoe is useful in cases where heel floating is used to correct hoof capsule distortion (Chapter10).

The section of fullered concave barstock usually used is $5/8"\times5/16"$ (16 x 8 mm) or $3/4"\times5/16"$ (20 x 8 mm), occasionally $5/8"\times1/4"$ (15 x 6 mm).

To calculate barstock length (Fig 16.2)

Measure from the point of the buttress across the apex of the frog to the toe for length (L); add this to the width of the foot (W); add the distance between the points of the heels (H); (and finally add $1^{1}/_{2}"$ (X). These calculations

CHAPTER 16

Figure 16.1:
Horseshoes can be made in a traditional coal or coke forge (seen here) or in a gas forge.

Figure 16.3:
Marking the barstock to show where to bend for the heartbar's frogplate.

Figure 16.4:
Forging the toe bend.

work well for feet in the 4½" to 5½" range. Larger feet may require (X) to be greater. W+ L + H + X = barstock length.

For a 5" × 5" foot with a heel width of 3" the barstock length would be 14½".

Forging

Mark the bar in the centre (Fig 16.3). If required the ends can be upset (thickened) for fire welding. The toe is turned (Figs 16.4 and 16.5). The scarf is prepared by knocking back the end (where not already upset) and thinning down (scarfing) (Fig 16.6). The heel and branch are turned and the nail-holes inserted by stamp and pritchel (Fig 16.7). The process is repeated for the other branch (Fig 16.8). The scarfs are brought together as a tight fit and fire welded (Fig 16.9). Up to this point the process is identical to the method used for making the eggbar shoe. The weld is now forged in to create the straight bar and frog plate (Fig 16.10 and 16.11). In order to fit the bar across the frog it may be necessary to set it away. If required, clip, rocker or roll the toe (Fig 16.12). To finish the shoe is boxed off as required (Fig 16.13).

Method of forging a heartbar shoe

Introduction

The heartbar shoe was first used, in modern times, by B. Chapman. It is formed in the shape of a heart and has an extended frog-plate finishing 1 cm from the apex of the frog. In recent years the heartbar shoe has come to the fore as the shoe to be used in the treatment of various forms of laminitic conditions. More recently many farriers have found it beneficial in the treatment of a variety of hoof wall lesions. A correctly fitted heartbar will reduce movement of the hoof capsule and can therefore be of use in cases of sand cracks and distal phalanx fractures.

A shoe, that can be described as a heartbar, is shown in Dollar and Wheatly (1881). In the book it is labelled as a fullered fore shoe (for harness horse) with frog plate (Fig 16.14). The heartbar is unusual but not unique as a shoe that has reappeared in recent times. The other that springs to mind is the eggbar shoe.

Well made, factory produced, heartbars are available (Brooks Lane Smithy, Cheshire, UK) although, at present, they are too cumbersome for the average Thoroughbred. Many farriers believe that the heartbar can only be fabricated in 2 pieces and then welded together. The author has found that any farrier, with average forging ability, can make a heartbar from one piece of barstock more efficiently than the fabricating method.

To calculate barstock length (Fig 16.2)

Measure from the point of the buttress to the apex of the frog (F), double that figure and reduce by one third (P). Add this to the width of the foot (W) plus the length (L) and finally add 2½" (X). Again, these calculations work well for feet in the 4½" to 5½" range. Larger feet require (X) to be greater. $F \times 4/3 = P$. P + W + L + X = bar length.

For a 5" × 5" foot with F = 3" the bar would be 16½", ie 4"+ 5"+ 5"+ 2½" = 16½".

Use ¾" × ⁵⁄₁₆" (20 × 8 mm) or ⅝" × ¼" (15 × 6 mm) fullered concave.

Forging

Mark the bar at both ends ½ of P (Fig 16.3). Draw out slightly and level the concave edge from the mark to the end of the bar. Turn the end to form the buttress and frog-plate (Fig 16.15). Repeat at the other end of the bar (Fig 16.16). Make the toe bend and branches. Stamp and pritchel the nail-holes and align the frog-plate in the centre of the branches (Fig 16.17). Turn the frog-plate 180° (Fig 16.18). Ensure that the 2 sides of the frog-plate align and butt together (Fig 16.19). Weld and draw down the frog-plate (Fig 16.20 and 16.21). Return the frog-plate 180° to its original position. If

CHAPTER 16

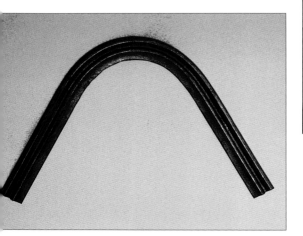

Figure 16.5:
The toe bend should be both central and symmetrical.

Figure 16.6:
The end of the bar is stepped down in preparation for welding. This is termed scarfing.

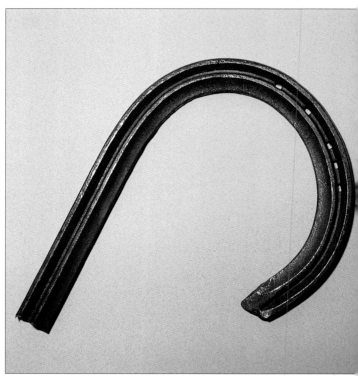

Figure 16.7:
The first branch has been turned, nail-holed and the scarf prepared.

required, rocker or roll the toe. Rasp the frog-plate, if necessary and box heels (Fig 16.22).

Method of forging an eggbar shoe

Introduction

The eggbar shoe is of ovoid shape, the bolder curve being at the toe (the ringshoe is similar being as wide posterior as anterior). The eggbar is an ideal shoe for use in the treatment of various forms of hoof imbalance. Its most common use is in the shoeing of dorso-palmar imbalances, especially where heels are under-run. The caudal support offered by the eggbar creates a biomechanical effect, moving the foot in relation to the limb towards the posterior (back under the leg). The biomechanics of using caudal support are explained in Chapters 1, 10, and 11. A correctly fitted eggbar will also reduce movement of the hoof capsule, ie shearing bulbs, and can therefore be of use in cases of medio-lateral imbalance.

The section of fullered concave barstock usually used is $5/8$" x $5/16$" (16 x 8 mm) or $3/4$" x $5/16$" (20 x 8 mm), or occasionally $5/8$" x $1/4$" (15 x 6 mm).

To calculate barstock length (Fig 16.2)

Measure from the point of the buttress across the apex of the frog to the toe for length (L); add this to the width of the foot (W); add the distance between the points of the heels (H); and finally add 2" (X). These calculations work well for feet in the $4^{1}/_{2}$" to $5^{1}/_{2}$" range. Larger feet may require (X) to be greater. W+ L + H + X = barstock length.

For a 5" x 5" foot with a heel width of 3" the barstock length would be 15".

Forging

Mark the bar in the centre. The method is idendical to the straight bar shoe until the welding is complete. If required the ends can be upset (thickened) for fire welding. The toe is turned (Figs 16.4 and 16.5). The scarf is prepared by knocking back the end (where not already upset) and thinning down (scarffing). The heel and branch are turned and the nail-holes inserted by stamp and pritchel (Fig 16.7). The process is repeated for the other branch. The scarfs are brought together as a tight fit (Figs 16.8 and 16.9) and fire welded (Fig 16.10). Care must be taken not to reduce the thickness of the shoe as this defeats the object of its use. Many horses, requiring eggbar shoeing, have hypertrophied (over developed) frogs. In order to fit the shoe it is often necessary to seat out the bar on the foot surface (Fig 16.23). If required, rocker or roll the toe (Fig 16.24). To finish the shoe, box off from quarter to quarter as required (Fig 16.23).

CHAPTER 16

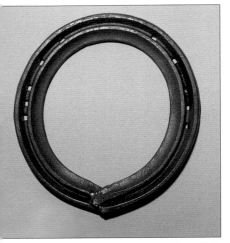

Figure 16.8:
The second branch is made and the scarfs brought together.

Figure 16.9:
Scarfs need to fit tightly prior to welding - like 2 hands clasped together.

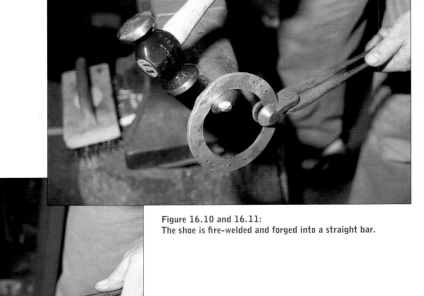

Figure 16.10 and 16.11:
The shoe is fire-welded and forged into a straight bar.

FORGING BAR SHOES

Figure 16.12:
A rolled toe is made by closing the fullering.

Figure 16.13:
The finished straight bar shoe.

CHAPTER 16

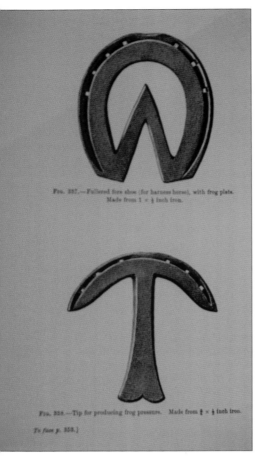

Figure 16.14:
A shoe, that can be described as a heartbar, is shown in Dollar and Wheatly (1881). In the book it is labelled as a fullered fore shoe (for harness horse) with frog plate.

Figure 16.15:
The bar end is drawn down and turned to form half the frog plate.

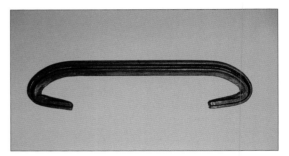

Figure 16.16:
It is important that each end is drawn down and turned the same.

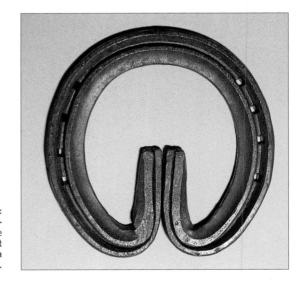

Figure 16.17:
The shoe is turned and nail-holed. The 2 sides of the frog plate are brought together to make a symmetrical shape.

162 FARRIERY – FOAL TO RACEHORSE

FORGING BAR SHOES

Figure 16.18:
The frog-plate is held in another pair of tongs and turned 180° prior to welding.

Figure 16.19:
If the sides of the frog-plate are not flush, tapping the end brings them together.

Figure 16.20:
The frog-plate is welded.

CHAPTER 16

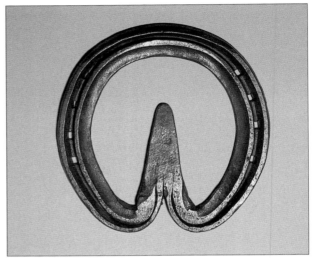

Figure 16.21:
The frog-plate is drawn down to the desired shape and length. The frogplate can be shortened if necessary by rasping.

Figure 16.22:
The finished heartbar shoe; the shoe has a rolled toe.

Figure 16.23:
The foot surface of the eggbar; the shoe has been seated out at the bar to accommodate a hypertrophied frog.

FORGING BAR SHOES

Figure 16.24:
A side view of the rocker toe on the eggbar shoe.

Figure 16.25:
The ground surface of the finished eggbar shoe.

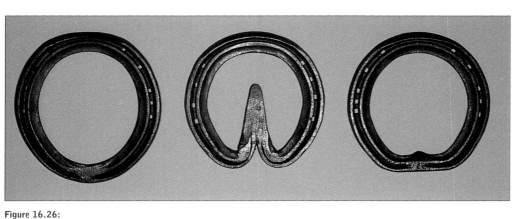

Figure 16.26:
The eggbar, heartbar and straightbar shoes.

CHAPTER 17

Terminology and technical language

This chapter classifies and clarifies many of the words and phrases used by farriers on the stud farm and racecourse. Terminology, used by farriers working with racehorses, varies between countries and even within them. Tools have been included in this list which is by no means comprehensive. Where the need has been felt, phrases and words have been attributed to either the UK or the USA. Fortunately, we are not quite the "two countries divided by a common language" as described by Winston Churchill.

Where words are used in definitions that appear elsewhere they are shown in italics. **Aka** is an abbreviation of *also known as*, additional words are listed. **See:** sends the reader to the author's preferred word. **See also:** signifies an additional linked word. An addition after words of either **(UK)** or **(USA)** indicates that the word is specifically from that country or at least clearly originates there.

Abaxial: Situated away from the centre.

Abscess: A localised infection of the sensitive tissues within the hoof. Abscesses often cause lameness which usually subsides when they are drained. **Aka**: Gravel, puss pocket.

Acetylene torch: A blowtorch which burns a combination of compressed acetylene gas and oxygen. Used for *welding*, *brazing* and cutting metals.

Acrylic: Any of numerous thermoplastic or thermosetting polymers or co-polymers of acrylic acid, methacrylic acid, esters of these acids, or acrylonitrile, used in horseshoeing to fill gaps or cracks in the hoof wall. **Aka**: Composite.

Acute: Coming about suddenly and severely, but persisting briefly.

Adhesion: Sticking together. The abnormal joining of living tissues.

Aluminium: (UK) A non-magnetic, conductive, metallic element; *racing plates* are made from an aluminium alloy.

Aluminum: (USA) **See:** Aluminium.

Ankylosis: From Greek angkylos, *crooked*. The fusing or consolidation of a joint.

Annular ligaments: *Ligaments* forming sheet-like bands to hold *tendons* in place.

Anterior: On or towards the front. **See also:** Posterior.

Antiseptic: Inhibits microbiotic growth.

Anvil: (Fig 17.1) *Anglo-Saxon* anfilt; *from* an, *on and* fealdan, *to fold*. A block of steel against which metals are hammered. The London pattern anvil, which began its evolution in the 13th century, has been the trademark of *farriers* and *blacksmiths* for 3 centuries. Shop anvils commonly weigh 123 kg and more.

Apprentice: Someone training in the workplace. In the UK there is an explicit agreement, in the US an apprenticeship is less formal.

Apron: Protective wear, usually leather, for the *farrier*. Covers upper legs and lower torso. Designs vary greatly from the traditional. **Aka:** Shoeing chaps, chaps.

Arc welder: A device which uses high-voltage electrodes to generate intense, concentrated heat. Used to weld and cut metal.

CHAPTER 17

Figure 17.1:
The Curtis Forge in Newmarket; the fires are traditional coke burning, the anvils are London pattern.

Arteriovenous anastomoses (AVA): Special blood vessels which act as bypass valves, diverting blood away from the tiny *capillaries* which nourish the *laminae*. Opening the AVA's reduces blood flow resistance, thereby increasing the flow rate. This is thought to be a mechanism for preventing frostbite in equine hooves.

Artery: A thick-walled vessel which carries oxygenated blood from the heart to the lungs and body.

Articular: Having to do with moving joints.

Articulation: The range of movement of a skeletal joint.

Atrophy: *From Greek* atrophia, *not to nourish*. Shrinking or degeneration of tissues. Usually results from disuse or disease.

AVA: See: arteriovenous anastomoses.

Axial: Situated toward the middle or centre.

Backwards shoe: (USA) (Fig 17.2) **See also:** Reverse shoe. A conventional horseshoe applied with the heels of the shoe at the toe of the hoof.

This usually requires that extra nail-holes be made. The backwards shoe acts as a rockered toe *eggbar shoe*. **Aka:** Reverse shoe; open toe eggbar; Napoleon shoe.

Bag of marbles: Multiple bone fractures in a small area.

Balance: A subjective term used in reference to both equine conformation and hoof geometry. Balance describes a condition in which each part is in optimum proportion to all others.

Bar shoe: Any *horseshoe* which is not interrupted by an opening between the heels. Various forms of bar shoe are used to increase support surface, apply pressure, prevent pressure, or stabilise the hoof. **See also:** Eggbar, straight bar, heartbar.

Barstock: The metal stock from which *horseshoes* are forged. Typical steel horseshoe barstock for racehorses in the UK ranges

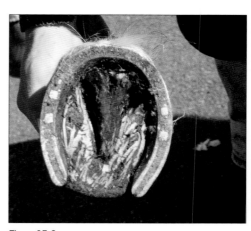

Figure 17.2:
A backwards shoe, also known as a reverse shoe and a Napoleon shoe.

168 FARRIERY – FOAL TO RACEHORSE

TERMINOLOGY AND TECHNICAL LANGUAGE

from ½" × ¼" and ⅝" × ¼" *(light-steel)* to ⅝" × ⁵⁄₁₆" and ¾" × ⁵⁄₁₆" *(training plates)*. In the USA *training plates* were traditionally *swedged*.

Basal crack: See: Grass crack.

Basement membrane: The delicate, microscopically thin layer of connective tissue between the secondary horny and secondary sensitive *laminae* within the equine *hoof*. The basement membrane is uniformly smooth and unbroken in healthy hooves, but breaks down with the onset of *laminitis*.

Bear foot: See: Clubfoot.

Bench knee: See: Offset knee.

Bifurcate: *From Latin* bi, *twice and* furca, *fork*. To separate, split or divide.

Big knee: See: Popped knee.

Bilateral: On both sides. Usually means both hooves of a pair.

Biotin: A colourless crystalline B complex vitamin C10H16N203S, essential for the activity of many enzyme systems and found in large quantities in liver, egg yolk, milk and yeast. Biotin is popularly believed to be beneficial to *hoof* growth and quality and is often included in horse feed supplements.

Blacksmith: A crafter of iron and steel. This term is sometimes inaccurately used as a synonym for *farrier*. Where farriers *forge* steel *horseshoes*, they are blacksmiths. But not all blacksmiths are farriers.

Blemish: A cosmetic flaw not affecting soundness.

Blocked heels: (USA) 1) The heels of a horseshoe which have been folded down against the ground surface of the shoe. This can be done to raise the heels or to act as a *heelcalk*. On thin shoes, the heel may be folded twice. 2) Large, square heel calks on manufactured racehorse shoes are sometimes called blocks, especially if the other heel is a *sticker*.

Blocking: (USA) Drawing up the nails, with *driving hammer* and *clenching block*, prior to *clenching*.

Borium: Grains or chips of tungsten carbide in a steel or brass matrix. Applied to the ground surface of a *horseshoe* in the *forge*, with an *acetylene torch*, or with an *arc welder*. Borium provides traction on hard, slick footing such as pavement. It also increases the wear life of the shoe. **Aka:** Tungsten (UK).

Bowed tendon: Damage or rupture of the fibres of a *tendon*, most often the SDFT of a foreleg. **Aka:** Tendinitis; peritendinitis; tendosynovitis; tendovaginitis.

Bow-legged: See: Carpal varus.

Boxed: Describes a *horseshoe* which has the outer edge of the hoof-facing side rounded or bevelled. This is done to prevent the exposed edge of the full-fit shoe from being pulled off should it be stepped on by another hoof's shoe.

Break-over: 1. The action of the hoof as it leaves the ground. 2. The point at which the foot leaves the ground.

Bruise: The rupturing of blood vessels within sensitive structures resulting from trauma. Hoof bruises often result from the horse stepping on stones. Bruises can also occur in any sensitive structure, including the frog and the bulbs of the heels. **Aka:** Strawberries.

Brushing: Interfering between paired hooves. **See also:** Interference.

Bucked shins: (USA) **See:** Shin splints.

Buffer: (UK) (Fig 17.3) Tool for cutting and/or raising *clenches*. **Aka:** Clinch cutter (USA).

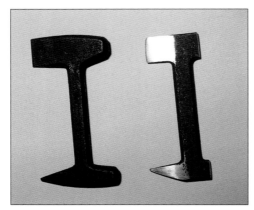

Figure 17.3:
Buffers (clinch cutters, USA) come in various sizes. Many platers remove clenches by rasping.

Bull-nosed: *Foot conformation* where the toe is over-long and very convex, the *HPA* is broken-back.

Burning on (UK): **See:** Hot setting.

Bursa: A sac of *synovia* located where friction would otherwise occur.

Bursitis: Inflammation of the bursa.

Buttress foot: See: Pyramidal disease.

By: Refers to a horse's paternal parentage. For example: Nashwan is by Blushing Groom.

Calk: *From Latin* calx, *the heel.* Any of several types of projection which may be forged on a *horseshoe*, welded or *brazed* on to a horseshoe, or inserted into a hole in the horseshoe. Calks are used to increase traction, alter movement, or adjust stance. **Aka:** Caulk: calkin

Calkin: See: Calk.

Cancellous: Loose bony tissue.

Canker: Abnormal, vegetative growth of the frog. May also affect the sole. **Aka:** Hoof cancer.

Cannon: The third metacarpal in the front leg, or the third metatarsal in the hind. **See:** Large metacarpal and large metatarsal.

Canted knee: A deviation similar in appearance to *offset knee*, but differing in that each row of carpal bones is *offset* from the bones above it, whereas offset knees have a relatively normal *carpus* with the metacarpals being displaced laterally at the carpo-metacarpal joint.

Capillaries: *From Latin* capillus, *hair.* The tiny, often microscopic, vessels which nourish the tissues and transfer blood from the *arteries* to the *veins*.

Capped hock: Subcutaneous *bursa;* bump under the skin at the point of the hock. This *blemish* may be of any size and is often caused by direct trauma to the hock.

Capped knees: See: Popped knee.

Capsular rotation: The hoof wall rotates away from the PIII, while the PIII remains in alignment with the pastern. This condition can be due not only to *founder* but also to overweight, too low a *hoof angle* or use of *toe-grabs*. **See also:** True rotation.

Carpal valgus: A conformation defect in which the fore limbs deviate *medially* above the knee, and *laterally* below the knee. This creates the appearance that the limbs are bent inward, under the horse. **Aka:** Knock-knees. **See also:** Valgus.

Carpal varus: A conformation defect in which the fore limbs deviate *laterally* above the knee, and *medially* below the knee. **Aka:** Bow-legged. **See also:** Varus.

Carpus: The equine knee, or human wrist - *carpal*, pertaining to the carpus.

Cartilage: Flexible, somewhat elastic, skeletal structures.

Cat walking: (USA) **See:** Plaiting.

Caudal: Rear surface. **See also:** Palmar and plantar.

Caudal extension: Where a shoe or other device continues horizontally beyond the *distal* border of the *hoof wall* towards the posterior. **See:** Extension, fishtail shoe.

Caulk: See: Calk.

Check ligament: *Ligament* which connects a tendon to a bone, to prevent hyperextension; check ligaments are often considered to be parts of tendons.

Chondro: Greek, having to do with *cartilage*.

Chondrocyte: A *cartilage* cell.

Chronic: *From Greek* Chronos, *time*. Persistent. The long-term phase of many diseases and conditions. **See also:** Acute.

Cleft: (USA) A horizontal crack in the *hoof* wall. Clefts are usually caused by damage to the coronary band or the rupture of an *abscess* at the coronary band. **Aka:** Cross crack and horizontal crack. **See also:** Sulci.

Clench: (UK) The part of a *horseshoe nail* visible on the outside of the shod hoof. This part of the nail is folded down against the hoof to form a clamp and normally has to be straightened or removed before the shoe can be pulled off without damaging the hoof. **Aka:** Clinch (USA).

Clencher: (UK) Tool for drawing down *clenches*. **Aka:** Clinchers (USA), clenching tongs (UK).

Clenching block: Tool used in drawing up the *nails*, prior to *clenching*.

Clinch (USA): **See:** Clench.

Clincher (USA): Tool for drawing down *clenches*. **See:** Clenchers.

Clinch cutter (USA): **See:** Buffer.

Clip: Flat projections, usually triangular or round, extending upward from the outer edge of a *horseshoe*. Clips are fit flat against, or set into, the outer surface of the *hoof* wall. Clips are used to prevent the shoe from shifting on the hoof, to stabilise the hoof wall.

Close nail: A *horseshoe nail* that does not actually *quick* the *hoof* but comes close enough to sensitive structures to create irritating pressure. It may take a few days for a close nail to cause the horse apparent discomfort. **Aka:** Hot nail (USA) and tight nail. **See:** Nailbind.

Clubfoot: An extremely upright hoof with a very broken-forward hoof-pastern axis. Caused by *flexor deformity*. In extreme cases, the *digit* may be folded back, with the animal bearing weight on its dorsal surface.

Coarse nail-hole: A nail-hole in a *horseshoe* which is located towards the inner edge of the shoe *web*. Toe nail-holes are often coarse punched. **See also:** Fine nail-hole.

Cold shoeing: (1. USA) To fit a horseshoe without hot-setting it. (2. UK) To prepare a horseshoe for shoeing without heat. **See:** Hot shoeing.

Collapsed heels: See: Under-run heels.

Collateral: Being of both sides, side by side.

Collateral cartilages: *Cartilage* structures which extend, one from either side, rearward from the wings of the PIII. These form the internal support *posterior* part of the hoof. **Aka:** Lateral cartilages.

Commissure: (USA) Groove on either side of the frog. **See:** Sulcus.

Common digital extensor tendon: A major *tendon* which runs down the dorsal surface of the leg, partially forms attachment at the *proximal phalanx* and joins with the branches of the suspensory ligament before inserting in the extensor process of the *distal phalanx.*

Composite: See: Acrylic.

Condyle: Articular prominence of the bone.

Conformation: The shape of the horse.

Congenital: A characteristic present at or soon after birth.

Contracted heels: See: Contracted hoof.

Contracted hoof: Condition in which the posterior half of the *hoof* undergoes a significant reduction in width. This may result from other *hoof* problems, improper shoeing or both. **Aka:** Contracted heels, hoofbound.

Contracted tendons: Term used to describe *flexor deformities* which cause heel-up *deformity* (DDFT) leading to *club foot* and *knuckling at fetlock* (SDFT). **See:** Flexor deformity.

Contralateral limb: Limb opposite the one that suffered the original lameness. Sometimes becomes lame from compensatory stress.

Contusion: *From Latin tundo, to beat.* A traumatic flesh injury which does not break the skin.

Coon foot: (USA) Very low *hoof angle* with an even lower pastern angle. May result from sprained suspensory *ligaments*, weak pastern or *chronic laminitis.*

Corium: See: Dermis.

Corks: A rural colloquialism which may refer to *heel calks, blocked heels* or turned down heels on *horseshoes.*

Corn: A bruise in the seat of corn sometimes caused by the heels growing over the shoe. Some conformations are predisposed to corns.

Corrective: 1) Trimming or shoeing a horse's hooves to counteract flaws in stance or gait. 2) Wrongly used as a synonym for *therapeutic.*

Cowboy shoeing: Horseshoeing by an incompetent shoer or careless shoeing performed by a *farrier.* This poor quality work may not be readily apparent to non-farriers. **Aka:** Shoe-horsing (USA); speed shoeing (USA).

Cranial: 1) The front surface of the limb. Towards the head. 2) Having to do with the cranium of the skull.

Crease: (USA) A groove cut into the ground surface of a *horseshoe* that has already been turned. Shoes are usually creased on the branches to provide a seat for the nail-heads. Creasing creates mild traction and allows the nails to be easily removed one at a time with crease nail pullers. **See:** Fullering

Crease nail pullers: (USA) **See:** Nail pullers.

Cross crack: See: Horizontal crack.

Crossfiring: A gait flaw which results in the collision of *diagonal* feet. This usually occurs at lateral gaits.

Cupping: The sole is trimmed out to a smooth convex surface.

Curb: 1) Swelling of the *plantar* surface of the hind leg just below the point of the hock may be caused by stress, poor conformation or direct trauma. 2) Sprain of the plantar *ligament*.

Cutters: (UK) **See:** Hoof cutters.

Dam: A horse's mother.

DDFT: See: Deep digital flexor tendon.

Deep digital flexor tendon (DDFT): A major *tendon* which runs down the back of the equine *leg*, uses the navicular bone as a fulcrum and inserts into the semilunar crest of the PIII.

Deformity: Any *conformation* characteristic that varies from the accepted ideal.

Dermis: The highly vascular connective tissue layer of the skin located below the *epidermis*, containing nerve endings, sweat and sebaceous glands and blood and lymph vessels. The sensitive *laminae* of the *hoof* are dermal. **Aka:** Corium.

De-rotation: 1) The distance between the outer surface of the *hoof* wall at the toe and the face of the PIII is greater near the coronet than it is near the ground. Often due to the *dubbing* of the toe. 2) Lowering the heel in cases of *flexor deformity* and *laminitis*.

Desmotomy: The surgical cutting of a *ligament*.

Diaphysis: The shaft of a long bone.

Digit: *From Latin* digitus, *a finger*. The equine limb distal to the fetlock.

Digital cushion: The sensitive, rubbery structure situated above the frog within the *hoof*. **Aka:** Plantar cushion.

Dimethyl sulfoxide: See: DMSO.

Dirt: Surface used for most *race tracks* when not on *turf*.

Dishing: (UK) A deviation in gait in which the *hoof* arcs inward (under the horse) in flight. Dishing is often seen in horses with *toed-out* conformation, and may lead to interference. **Aka:** Winging-in (USA).

Distal: When referring to limbs, distal means away from the torso or comparatively farther from the torso. Opposite of proximal.

Distal interphalangeal articulation (coffin joint): The joint that includes the *middle phalanx*, the *distal phalanx* and the *distal sesamoid*.

Distal phalanx: The most *distal* bone in each equine *limb*. It is situated completely within the *hoof* and resembles the hoof in basic shape. **Aka:** Coffin bone; distal phalange; pedal bone; PIII ; third phalanx; os pedis.

Distal sesamoid: Small bone within the distal interphalangeal articulation (coffin joint) distal to the *middle phalanx* and posterior to the *distal phalanx*. **Aka:** Navicular bone.

Dorsal: *From Latin* dorsum, *the back*. 1) The front surface of the equine *hoof* and *leg*. 2) When referring to the entire animal, dorsal means the spine or centre-line of the back.

Dorsal wall resection: Removal of part or all of the dorsal hoof wall, usually as part of treatment for *founder*.

Dorso-palmar (D-P): Imaginary line running from front to back obliquely; viewing the *leg* from the side shows dorso-palmar *balance*.

Drawing knife: (UK) (Fig 17.4) *Farrier's* knife with a curved blade for trimming the *hoof*. **Aka:** Hoof knife (USA).

Dressing: See: Trimming.

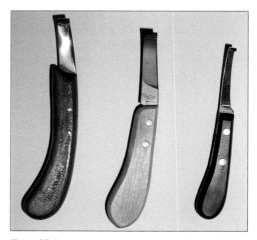

Figure 17.4:
Drawing knives (also known as hoof knives) come in various sizes. The thinner bladed knife (right) is a searcher, platers often use these as their everyday knife.

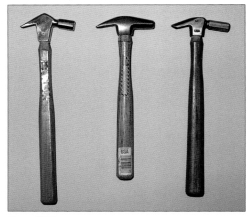

Figure 17.5:
Driving hammers for plating are usually 30–36 g (10–12 oz).

Driving hammer: (Fig 17.5) The claw hammer that is used for *nailing racing plates* on. **Aka:** Nailing-on hammer.

Dropped sole: The sole of a hoof which has become convex rather than concave. As the sole protrudes below the *solar* plane of the *hoof* wall, it bears excessive weight and is subject to *bruising*. Dropped sole is a *laminitic* condition.

Dubbed toe: (USA) 1) A hoof which has had the *dorsal* surface of its toe ground off. This may be the result of excessive rasping after a *horseshoe* was poorly fitted or of high *lameness* which may cause the horse to drag his *hoof*. 2) The intentional dressing back of the toe done to treat *founder* or a toe *flare*. **Aka:** Dumped toe (UK).

Dumped toe: (UK) **See:** Dubbed toe.

Eggbar: A *shoe* of ovoid shape, used for dorso-palmar *imbalances*.

Epidermis: The outer, protective, nonvascular layer of the skin which covers the *dermis*. The *hoof* wall, horny *laminae* and other horny hoof structures are epidermal.

Epiphyseal plates: *Cartilages* near the ends of bones which allow them to grow lengthways. As the horse matures, the plates *ossify* or close. The more distal plates close first. **Aka:** Growth plates; epiphyseal cartilages and physes.

Epithelium: Thin membrane tissues covering most of the body's structures and organs, internal and external. Also describes the first layers that heal over a wound.

Etiology: The study of causes. **Aka:** Aetiology.

Exfoliate: To shed or flake off dead tissue. The sole of the *hoof* for example, normally exfoliates as it grows down.

Exostosis: Abnormal bony growth. New growth protruding from the surface, eg *shin splints*.

Expansion: 1) The very slight outward movement of the quarters of the *hoof* which may occur during weightbearing and/or the increase in *solar* width which occurs as the hoof grows down. 2) In reference to a *horseshoe*, expansion describes the practice of fitting the *posterior* half of the shoe larger than the hoof to allow for hoof expansion.

Extension: Where a shoe or other device continues horizontally beyond the *distal* border of the *hoof wall*. **See:** Lateral extension, toe extension, caudal extension.

Extensor process: The point of insertion of the main digital extensor tendon into the PIII.

Extensor process disease: See: Pyramidal disease.

False quarter: A lesion in the *hoof wall* parallel to the *horn* tubules, resulting from an injury to the coronary band.

False sole: See: Retained sole.

Farrier: One who applies horseshoes to equine and/or bovine animals.

FD: **See:** Flexural deformity.

Feral: Animals of domestic ancestry who have reverted to the wild state. New Forest ponies and American mustangs are feral, rather than truly wild.

Fever rings: (USA) **See:** Hoof rings.

Filly: A female horse under 4-years-old.

File: See: Rasp.

Fine stamped: A nail-hole in a *horseshoe* which is located relatively close to the outer edge of the *web*. Heel nails are often punched fine. **See also:** Coarse punched. **Aka:** Fine punched.

Fishtail shoe: (Fig 17.6) *Horseshoe*, instituted by the author, extending to the *anterior* almost to the fetlock for ultimate *caudal* support. **See:** Caudal extension.

Fitting hammer: The hammer of 1–2 lb ($1/2$–1 kg) used with a *stalljack* to adjust the shape of the shoe.

Flaccid tendon: Weak flexor tendons in a young *foal* causing a toe-up *conformation*.

Flare: An outward distortion which may occur on any portion of the *hoof* wall. If left untreated, they can alter functional *toe angle, medio-lateral balance* and hoof symmetry.

Flexion: Decreasing the angle of the joint.

Flexural deformity: *Dorso-palmar imbalance* involving flexor tendons creating severe broken-back and/or broken-forward *HPA*. Conditions resulting from flexural deformity include *contracted tendon, club foot, flaccid tendon knuckling at fetlock*.

Flipper foot: An extremely overgrown, *toe-flared hoof*. Can result from *founder* or neglect. In severe cases, the horse may stand on the back of its pastern, allowing the *solar* aspect of the *hoof* to be seen from in front of the animal.

Floating the heel: A section of the *hoof wall* from medial quarter to heel is removed. When the *shoe* is then attached there is a clear gap between *wall* and *shoe*.

Foal: A very young horse of either sex. A pregnant *mare* is said to be 'in foal'. The birth process of horses is referred to as 'foaling'.

Foot: The *hoof* and all that is contained within it.

Foot rest: Tool for taking the weight of the foot when pulled forward, often a tripod. **Aka:** Leg stand.

Forge: 1) A furnace used to heat metal. 2) The workshop of a farrier or blacksmith. 3) To make something out of metal.

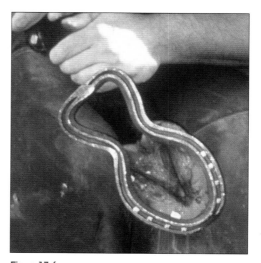

Figure 17.6:
Fishtail shoe; this was developed by the author to give ultimate caudal support. It is useful in cases of flexor tendon injury and after the removal of a cast in some compound fracture cases.

Forging: A fault in gait which results in the toe of a hind *hoof* striking the *solar* surface of its *lateral* fore hoof. Similar, but not identical, to *over-reaching*.

Founder: *Old French* afondrer, *to founder or sink. From Latin* fundus, *bottom*. The mechanical result of *laminitis*. The tip of the PIII sinks downward towards the sole near the toe. *Sole bruising, abscesses* and *hoof* distortion commonly occur.

Founder stance: The standing position often assumed by horses during *acute laminitis*. The hind feet will be placed far forward of their usual position and will bear an inordinate amount of weight. The fore hooves will be placed out in front of the animal and may bear weight only at the heels. **Aka:** Laminitic stance.

Founder treatment, Chapman: Features the support of the PIII via carefully applied frog pressure. Supporting the PIII in this way prevents further sinking of the bone and restores the circulation of blood which was compromised by the founder. This is often accomplished with a *heartbar shoe*. Popularised by B. Chapman.

Founder treatment, classic or traditional: Features lowering the heels in foundered hooves and dressing back the toes as far as reasonably possible. Lowering the heels aids the horse in his natural reaction, which is to shift weightbearing away from the damaged toes and onto the heels. It also causes the frog to bear weight, which provides some mechanical support for the PIII. Dressing back the toe eases break-over and reduces the peeling stress which could damage surviving *laminae*.

Founder treatment, Redden: Features raising the heels of foundered hooves to reduce the tension of the DDFT on the PIII. The pull of the DDFT is considered to be a major cause of the downward rotation of the PIII. Popularised by R.F. Redden.

Four point trim: A hoof trimming technique (instigator: G. Ovnicek) in which the heels are trimmed back to the widest point of the frog, then the toe is bevelled in a manner akin to what would be done in preparation for a *rockered toe horseshoe*. The quarters are then rasped until they would no longer bear weight on a firm surface. This leaves only 4 full-loading points on the hoof wall; one at each side of the toe and one at each heel. This method is based on observations of the hooves of feral horses and its advocates claim that it results in the development of stronger hoof structure.

Full pad: Layer of material, usually leather or plastic between the *shoe* and the wall, covering the sole and frog. **See:** Pad, rim pad.

Gelding: A castrated male horse of any age.

Get: A collective term for all of a given horse's offspring.

Glue-on shoe: A *horseshoe* that is attached by adhesive rather than by nailing. The 2 most common are the Mustad Race-glu and the Dalric cuff.

Grass crack: A split in the hoof wall running parallel to the horn tubules and originating at the distal wall. It may be superficial or deep. **Aka:** Basal crack.

Gravel: See: Abscess.

Grooving: (1. USA) Cutting or burning a horizontal channel across the fibres of the *hoof horn* to alter the way in which stresses are transferred up the wall. This may be done when treating *founder, flares, basal cracks* and other hoof maladies. (2. UK) Cutting or burning one or more channels either parallel to the horn tubules, vertical to the hoof or horizontal to the hoof.

Growth plates: See: Epiphyseal plates.

TERMINOLOGY AND TECHNICAL LANGUAGE

Growth rings: Roughly horizontal distortions on the hoof wall which may be caused by changes in the diet, environment, season or by illness. Uneven hoof rings may indicate that the horse has been *foundered*. **Aka:** Hoof rings: fever rings.

Half-round cutters: Type of *hoof cutters* for removing *dorsal hoof wall*.

Hammer: See: Driving hammer, fitting hammer.

Heel calk (USA): A projection forged or welded onto the ground side of the heel of a horseshoe. Heel calks provide braking traction as the hoof lands, but no grip at break-over.

Heartbar: A *shoe* instituted by B. Chapman, in the shape of a heart, that has an extended frog-plate finishing 1 cm from the apex of the frog. Used for *laminitis* and other *hoof wall lesions*.

Hoof: (1. UK) The horny covering of the foot. It includes the periople, hoof wall, white line, sole and frog. (2. USA) The equine foot, includes the coronary band and all parts *distal*. Sometimes refers to only the horny parts of the foot. **See also:** Foot.

Hoof angle: The angle at which the *dorsal* line of the hoof intersects with the plane of its *solar* surface. Hoof angle can be measured with a tool called a hoof gauge or hoof protractor. **Aka:** Toe angle.

Hoofbound: See: Contracted hoof.

Hoof cancer: See: Canker.

Hoof horn: The tough, insensitive parts of the hoof, such as the wall, are made of horn. The wall is composed of fibres which grow downward from the coronary band called tubular horn. These are cemented together by intertubular horn. The approximate moisture content of hoof horn is 25% for the wall, 33% for the sole and 50% for the frog.

Hoof knife: (USA) **See:** Drawing knife (Fig 17.4).

Hoof nippers: (USA) (Fig 17.7) **See:** Hoof cutters.

Hoof-pastern axis (HPA): The alignment of the *hoof capsule (dorsal hoof wall)* with the pasterns. Hoof-pastern axis can be said to be straight, broken-back and broken-forward.

Hoof cutters: (UK) (Fig 17.7) Tool for removing *hoof wall* during *trimming*, most *platers* use size 12" racetrack. **Aka:** Hoof nippers (USA), nippers, cutters, croppers (UK). **See also:** Half-round cutters.

Hoof rings: See: Growth rings.

Hoof sealant: Any of a number of artificial varnishes which reduce the transfer of moisture between the hoof and the environment.

Hoof tester: (Fig 17.8) A tool used to squeeze the foot to locate an area of pain.

Horse nail: See: Nail, horseshoe.

Horseshoe: A device affixed to the equine hoof to protect it from wear and damage, provide grip and occasionally for therapeutic reasons. Most horseshoes are made of *steel (training*

Figure 17.7:
Shoeing tools are made with surgical precision. The types used by farriers working on Thoroughbreds are usually the lightest offering the most finesse. From left to right; pincers (shoe pullers, USA), hoof cutters, clenchers (clinchers USA), nail pullers.

CHAPTER 17

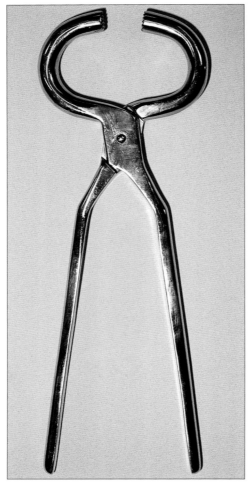

Figure 17.8:
Hoof tester; a tool used to squeeze the foot to locate an area of pain.

plates) or *aluminium (racing plates)* and affixed with nails driven through the *hoof wall*, but there are some types which are made of other materials and attached in other ways *(glue shoes)*.

Horseshoe nail: See: Nail, horseshoe.

Horseshoer: (USA) Any person who applies *horseshoes* to horses' *hooves*. This term does not imply any particular level of skill. **See also:** Farrier, plater.

Hospital plate shoe: *Horseshoe* with a removable plate, *aluminium* or *steel*, for treating *lesion* of the *foot*.

Hot nail: (USA) **See:** Nail bind.

Hot shoeing: (1. USA) Horseshoeing with the aid of a forge. May involve fabricating the *horseshoes* from *barstock* and/or *hot setting*. (2. UK) Fitting a horseshoe by *hot setting*, ie *burning on*.

HPA: See: Hoof-pastern axis.

Hyperextension: A joint extended beyond its normal range.

Hyperflexion: A joint flexed beyond its normal range.

Hypertrophy: Increase in tissue volume resulting from the enlargement of existing cells.

Imbalance: Where the foot and/or limb are not considered to be in *harmony*. **See:** Balance.

Interdigitate: To fold or lock together, as when the fingers of one hand are laced between those of the other. This term describes the manner in which the horny and sensitive *laminae* interlock within the *hoof*.

Interfering: 1) Generally describes any *hoof* which strikes another of the horse's *legs* in movement as a result of a fault in gait. 2) Specifically, collision between paired legs. **See:** Brushing, cross-firing, forging, over-reaching, scalping.

Iron: A metallic, electroconductive element. Wrought iron was a common form of impure iron with a fibrous structure. It was workable cold, resisted burning and forge-welded easily. *Mild steel*, which has replaced wrought iron in general use, is often called iron. **See also:** Mild steel.

Ischaemia: (UK) Lack of oxygenated blood flow to the tissues. **Aka:** Ischemia (USA).

IUJH: (USA) International Union of Journeyman Horseshoers. A *farrier* trade union started in 1874 as the Journeyman Horseshoers National Union. The name was changed to its current form in 1893. In 1934 the primary emphasis of the IUJH shifted from draft horses to track horse shoeing. **Aka:** JHU; Platers Union.

Jammed heel: (USA) One heel appears to be jammed up into the foot, with its bulb and coronary band correspondingly distorted. Often wrongly called sheared heels. **Aka:** Heel shunt.

Keg shoe: (USA) A conventional, factory made *horseshoe*.

Keratin: *From Greek* keratos, *horn*. The tough protein component in horn, hair, skin and hooves.

Keratoma: A tumour of the horny *laminae*. Often seen as an inward distortion of the *white line*. **Aka:** Keraphyllocele.

Knee-hitting: Interference in which the fore hoof strikes the inside of the knee of the opposite limb.

Knife: See: Hoof knife, drawing knife, loop knife.

Knock-knees: See: Carpal valgus.

Knuckling at fetlock: *Conformation* fault where the pastern is upright due to a *flexural deformity* involving the SDFT.

Lame: Adjective describing a horse who is suffering sufficient pain and/or mechanical defect to interfere with normal movement and weightbearing in one or more *limbs*. Limping.

Lameness: Any limping affecting natural gait or stance, not necessarily an injury. Graded on a scale of 1 to 5; 1 being very subtle, 5 being non weightbearing.

Lamina (plural laminae): The tissues which attach the PIII to the hoof wall. The inner laminae are attached to the bone and are called sensitive laminae. The outer laminae are attached to the wall and are called horny laminae.

Laminitic: Adjective describing any condition involving *laminitis*.

Laminitis: Strictly an inflammation of the laminae but in the horse a systemic illness which involves the malfunction of the AVAs in the hooves. The blood flow through the hoof may be increased, but it is diverted away from the fine *capillaries* which supply the *laminae*. This results in the death of some laminar tissue and causes the horse pain. Laminitis can lead to *founder*.

Large metacarpal: Bone distal to the *carpus* and proximal to the *fetlock*. **Aka:** Cannon, third metacarpal, MIII.

Large metatarsal: Bone distal to the *tarsus* and proximal to the *fetlock*. **Aka:** Cannon, third metatarsal, shannon, MIII.

Lateral: 1) Outside or away from the body's centre line. Opposite of *medial*. 2) On the same side of the horse; such as the *near* foreleg and the near hind. 3) Towards or on the side of something.

Lateral cartilages: See: Collateral cartilages.

Lateral digital extensor tendon: A major *tendon* which runs down the dorso-lateral surface of the leg, the dorsal surface of the fetlock, before inserting in the *proximal phalanx*.

Lateral extension: 1) Generic term for all types of *shoes*, or other devices, that continues horizontally beyond the *distal* border of the *hoof wall*, on either side. 2) Specifically a shoe, or

CHAPTER 17

other device, that continues horizontally beyond the *distal* border of the *hoof wall*, on the lateral side. **See:** Medial extension.

Layup: (USA) A racehorse taking a break from the racetrack either for recuperation or due to injury.

Leg: The equine *limb* from the knee or hock down.

Lesion: Site of structural change in body tissue, usually due to injury.

LHLT: Low-heel, long-toe: A condition in equine hooves in which the heels are excessively low due to trimming, wear or *under-run* growth; and the toe is excessively long, often in the form of a *flare*. LHLT is known to contribute to *navicular disease* and gait defects such as *forging* and *over-reaching*. **Aka:** LTLH

Ligament: Ligaments are strong fibrous tissues which connect bones to one another (except for *check ligaments*, which connect *tendon* to bone). Ligaments also hold tendons in place, eg palmar annular ligament. Ligaments are subject to sprains and tears.

Light steel: (UK) Term describing *steel barstock* sizes used on *racehorses* in the UK. Usually either $1/2" \times 1/4"$ or $5/8" \times 1/4"$ (12 x 6 or 15 x 6 mm).

Limb: The entire equine appendage, from the scapula or hip down.

Loop knife: (Fig 17.9) Type of farriers knife where the blade is an oval with an eye, forming 2 blades.

Luxate From Greek *loxos, slanting*. To put out of joint or dislocate.

Mare: A female horse, usually 4 years or older.

Master: 1) Traditionally a tradesman who is proprietor of his own business. Especially if that business employs others. 2) A trainer of *apprentices*.

Medial: Inside or toward the body's centre line. **See also:** Lateral.

Medio-lateral balance: Describes the *conformation* of the *limb* and *foot* from an *anterior* and/or *posterior* view-point.

Middle phalanx: Bone distal to the proximal phalanx and proximal to the distal phalanx **Aka:** Short pastern, second phalanx, PII.

Mild steel: Low carbon *steel* used in the manufacture of *horseshoes* and *barstock*.

Mustad: Swedish horseshoe nail manufacturer since 1885, also produces Race-glus.

Figure 17.9:
A loop knife; a type of farriers knife where the blade is an oval with an eye, forming 2 blades. The smaller loop knife on the left is often called an abscess loop or searcher.

FARRIERY – FOAL TO RACEHORSE

Nail, horseshoe: (Fig 17.10) Soft *steel* nails specially designed for attaching *horseshoes* to hooves. They are generally made with a 4-sided, tapered shaft; the tip bevelled on the inside and a head shaped to seat into a horseshoe. A pattern or trademark is stamped into the inside face of the head to make it possible to distinguish the inside from the outside at a glance. **Aka:** Horse nails, plate nails.

Nailing: Process of driving nails through a *shoe* into the *hoof*.

Nailing-up: Replacing some *nails* without removing the *shoe* from the *hoof*.

Nail pullers: (UK) Tool for removing individual nails. **Aka:** Crease nail pullers (USA).

Napoleon shoe See: Reverse shoe.

Navicular bone: *From Latin* navicula, *little ship*. **See:** Distal sesamoid.

Near: The horse's left side. The side of the horse that faces west when the animal is walking north. Horses are most often led, saddled and mounted from the near side. Opposite of *off*.

Necrosis: Death of tissue.

Off: The horse's right side. The side of the horse that faces east when the animal is walking north. Opposite of *near*.

Offset knee: A limb conformation defect in which the *leg* fits somewhat to the outside of, rather than directly below, the forearm and the knee. **Aka:** Bench knee.

Onychomycosis: Means a fungal disease of the nail or claw. Denotes the decay of the inner *hoof* wall and the *white line* as a result of infection by highly adaptable micro-organisms. **Aka:** White line disease; onycholysis; seedy toe.

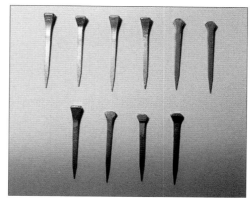

Figure 17.10:
Nails come in a variety of sizes for both plates and shoes. Training shoes in the UK usually take E head size 2 and 3, racing plates take plate nails with city heads. In America racehorses are shod and plated in Race Nails (RN) sizes 3 to $4^1/_2$, depending on shoe type and the farrier's choice.

Open-toed eggbar: See: Reverse shoe.

Opening the heels: Trimming away the bars and some of the heel buttress to give the appearance of wider heels.

Osselets: Arthritis of the fetlock joint, characterised by hard swelling.

Ossification: The hardening of soft tissues, such as *ligament* or *cartilage*, into bone. This is often a part of the natural ageing process.

Osteolysis: Degeneration of bone.

Out of: Refers to the horse's maternal parentage. For example: Nashwan is out of Height of Fashion.

Over-reach: An injury to the bulb of the heel due to *over-reaching*.

Over-reaching: A fault in gait which causes the toe of a hind *hoof* to strike the back of the *lateral* fore *leg* or the heel of its *hoof* or *horseshoe*. Similar, but not identical, to *forging*.

PIII: See: Distal phalanx.

PIII rotation: See: True PIII rotation.

Pad: Layer of material, usually leather, plastic or metal, between the *shoe* and the sole. **See:** Full pad, rim pad.

Paddling: A deviation in gait in which the *hoof* arcs outward in flight. Paddling is often seen in horses which have a *toed-in* conformation and, while although it is more obvious to the untrained eye than *dishing*, it is less likely to cause *interference*. **Aka:** Winging-out (USA).

Paddock farrier: (USA) *Farrier* employed by *racetrack* during *racing*.

Pair: Two *horseshoes* or hooves. Unless otherwise specified, a pair means either both fronts or both hinds.

Palmar: Rear surface of lower limb.

Palmar process: The rearmost portion of either side of a PIII. This is where the *lateral cartilages* attach to the bone. **Aka:** Volar

Palpate: To examine by touch.

Patch: A repaired *hoof crack*. **See:** Patching.

Patching: 1) Repair of a *hoof crack*, usually a deep coronary band *quarter crack*. 2) There are many ways to patch a *crack*. The same principals are basic to all successful methods. The *patch* must be strong, safe to apply and durable.

Patella: One of the bones forming the stifle joint and a human knee.

Patho- or path-: *Greek* pathos. Prefix denoting disease or suffering.

Pathogenesis: Origin of suffering. The generation and development of a disease.

Pathological: See: Therapeutic.

Pathology: The study of alterations produced by disease.

Patten shoe: A form of *bar shoe* that elevates the heels and breaks forward the *HPA*. Used to treat *flexor tendon* injuries. **Aka:** Rest shoe.

Pedal osteitis: Severe and/or repeated *bruising* of the sole resulting in the inflammation of the *distal phalanx*.

Pelvic limb: A hind *limb*.

Penetrating crack: Any kind of hoof crack which exposes sensitive tissue and/or causes *lameness*. **Aka:** Deep crack.

Peri-: *Greek*. Prefix meaning around or enclosing.

Periarticular: Situated around a joint.

Periople: The thin, tough, protective covering of the coronary band. The periople normally extends less than one inch down the hoof wall.

Periosteum: The fibrous membranes which cover the bones.

Periostitis: See: Exostosis.

Peritendinitis: See: Bowed tendon.

Phalange: See: Phalanx.

Phalanx: Any of the major bones in a *digit*. Plural phalanges.

Physiology: Study of the function of parts of the body.

Pigeon-toed: See: Toed-in.

Pincers: (UK) Tool for removing *shoes*. **Aka:** Shoe pullers (USA).

Plaiting: (UK) Placing one foot in a pair, in front of the other. Horses that plait are likely to *interfere*. **Aka:** Rope walking, cat walking and wolf walking.

Plantar: *From Latin* planta, *the sole of the human foot.* The back side of the horse's hind leg.

Plantar cushion: See: digital cushion.

Plate: See: Racing plate.

Plate nails: Special small city head nails that fit *racing plates* and *light steel*. UK sizes are ASV $1^5/_8$" and ASV $1^3/_4$", US sizes are RN 3 and RN $3^1/_2$.

Plater: A *farrier* who specialises in shoeing race horses.

Platers Union: See: IUJH.

Plating: Shoeing race horses.

Plexus: *Latin.* A network of nerves or blood vessels.

Popped ankle: (USA) **See:** Windpuff.

Popped knee: (USA) Any of several forms of inflammation of the *carpus*. May be a bursitis, a herniated joint capsule, or a distended *tendon* sheath. **Aka:** Big knee; capped knee; hygroma.

Popped sesamoid: (USA) Inflammation of a proximal sesamoid bone or a sesamoidian *ligament*. May result from uneven stress on the fetlock, or from direct injury such as may be caused by interference. **Aka:** Sesamoiditis (UK).

Posterior: Towards or on the back surface. Opposite of *anterior*.

Prick: To drive a nail into the sensitive tissue. **Aka:** Quick; hot nail.

Profile: Cross-section through a *horseshoe* or *barstock*. **Aka:** Section.

Prolapsed sole: See: Dropped sole.

Proximal: In reference to *limbs*, proximal means close to the torso, or comparatively closer to the torso. Opposite of *distal*.

Proximal phalanx: Bone distal to the large metacarpal and proximal to the middle phalanx. **Aka:** Long pastern, first phalanx, PI.

Pus pocket: See: Abscess.

Pyramidal disease: Severe inflammation of the PIII at the extensor process where the main digital extensor *tendon* is attached. *Dorsal* swelling above the coronary band may occur. **Aka:** Buttress foot; extensor process disease.

Pyramidal process: See: Extensor process.

Quarter crack: 1) Any *sand crack* in the quarters of the *hoof* wall. May be *superficial* or *penetrating*: *basal* or *coronary*. 2) This term has been used to specifically denote a coronary sand crack in the quarter to heel. **See also:** patching.

Quick: *From Anglo-Saxon* cwic, *alive or living.* 1) Any of the sensitive structures within the *hoof* capsule. 2) To penetrate the sensitive tissue of a foot. **See:** Prick.

Quittor: (Fig 17.11) *From Old French* quitture, *discharge.* Necrosis of the *lateral cartilages* due to infection. Characterised by severe *lameness* and pus discharge. A *laminal abscess* existing at the *coronary band* is often mistakenly described as a quittor.

Racecourse: (UK) Venue for *racing*.

Racecourse farrier: (UK) *Farrier* employed by the *racecourse* on racing days. **See:** Paddock farrier (USA).

Racehorse: Any horse used in organised racing, usually a *Thoroughbred*. *Arabian* racing has become popular in recent years.

CHAPTER 17

Figure 17.11:
A quittor is a necrosis of the lateral cartilages due to infection. Characterised by severe lameness and pus discharge from above the coronary band. A laminal abscess exiting at the coronary band is often mistakenly described as a quittor.

Racetrack (USA): Venue for *racing*.

Racing plate: A horseshoe made of aluminium for racing. Many racehorses remain in racing plates throughout training. **Aka:** Plate.

Radiograph: (Fig 17.12) An image produced by photographing artificially generated radiation which passes through visually opaque matter. Only dense objects, such as bone, are normally visible on radiographs. **See also:** X-ray.

Rasp: (Fig 17.13) Tool for *trimming* the *hoof wall*. Usually a 12", tanged, 'Platers rasp'. 'New rasp' used only on the *foot*, 'old rasp' used for *clenching* and *rasping shoe*.

Rasping: *Trimming* the *hoof wall* with a *rasp*.

Refit: (UK) **See:** Reset.

Relieved: (USA) **See:** Seated.

Relieving: (USA) **See:** Seating.

Remove: (UK) **See:** Reset.

Resection: An operation involving the removal of part of an organ or structure. **See:** Dorsal wall resection.

Reset: (USA) To remove a *horseshoe*, trim the *hoof*, then reattach the same horseshoe. **Aka:** Shift, refit, remove.

Retained sole: The sole of a *hoof* when it does not *exfoliate* normally. **Aka:** False sole.

Reverse shoe: (UK) A conventional horseshoe applied with the heels of the shoe at the toe of the hoof. This usually requires that extra nail-holes be made. The reverse shoe acts as a rockered toe *eggbar shoe*. **Aka:** Backwards shoe; open toe eggbar; Napoleon shoe.

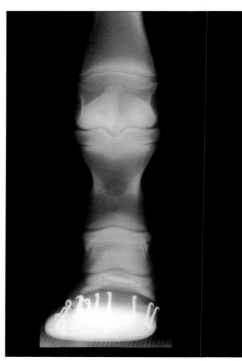

Figure 17.12:
A Radiograph is an image produced by photographing artificially generated radiation which passes through visually opaque matter. Only dense objects, such as bone, are normally visible on radiographs. The image shown is of a 2-month-old foal. The epiphyseal cartilages (growth plates) can be seen. The growth plate in the proximal phalanx is partially closed and the one at the distal large metacarpal is still open. The foal has a nailed on medial extension which shows clearly.

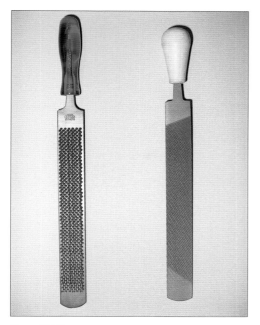

Figure 17.13:
A platers rasp; it is otherwise described as a 12" tanged rasp, it has a rough side (left) and a smooth side (right).

Rim Pad: Layer of material, usually leather or plastic between the *shoe* and the wall, not covering the sole. **See:** Pad, full pad.

Ringbone: *Exostosis* in or around the coffin or pastern joint, or on the pastern bones. High ringbone is in or near the pastern joint and is usually palpable. Low ringbone is in or near the coffin joint, and often is not directly visible. Articular ringbone actually involves a joint. Periarticular ringbone is located around, but not within a joint. Exostosis on the pastern bones, between the joints is called false ringbone or non-articular ringbone.

Road puff: (USA) **See:** Windpuff.

Road work: (UK) Term used to describe exercise on public roads used to build stamina and condition legs.

Rockered toe: A *horseshoe* that has been curved upward toward the hoof at the toe. This eases and directs *break-over*. Hooves must be specially prepared to receive rocker toe shoes. **Aka:** Rolled toes (UK).

Rolled toe: A *horseshoe* that has been rounded or bevelled on the outer edge of the ground surface at the toe. This eases *break-over*. The hoof side of the shoe is left flat, so the hoof needs no special preparation.

Rope walking: (USA) **See:** Plaiting.

Rotational deformity: A medio-lateral deformity where all or part of the limb turns outward along a horizontal plane. **See also:** True PIII rotation.

Run-under heels: See: Under-run heels.

Saber-legged: (USA) **See:** Sickle-hocked.

Safed: Describes a *horseshoe* which has the outer edge of the ground surface of the medial branch rounded. This is done to decrease the chance of it pulling another shoe or causing injury should the horse *interfere*.

Sand crack: A hoof crack parallel to the *horn* tubules, originating at the coronary band. May be *superficial* or *penetrating* and can occur anywhere in the hoof wall.

Scalping: Fault of gait which results in the toe of a fore hoof striking the *dorsal* surface of the *lateral* hind *hoof* or *leg*.

SDFT: See: Superficial digital flexor tendon.

Sealant: See: Hoof sealant.

Seated: A *horseshoe* that is sloping or depressed on the inner half of the *web* on the foot surface, except for the heels. This is done to prevent the shoe from applying pressure on the sole of the *hoof*. **Aka:** Relieved.

Seating out: The process of creating a seated shoe. **Aka:** Relieving.

Section: See: Profile.

Sepsis: The presence of disease-causing organisms or their toxins in the blood or tissues.

Sequestrum: *From Latin* sequesto, *to sever*. Portion of bone which has become detached in *necrosis*.

Sesamoiditis: 1) Inflammation of the proximal sesamoid bone(s). 2) May involve an actual fracture of a sesamoid bone. Sesamoiditis can be the result of direct injury, uneven weightbearing or fatigue. **Aka:** Popped sesamoid.

Set: Four *horseshoes*.

Set toe: The toe of the shoe is bent up at 90° - usually a hind shoe.

Shannon: See: Large metatarsal.

Sheared heels: Failure of internal structures which normally bind the heels together. Allows the heels to move independently more than normal and can cause *lameness*. This term is sometimes used erroneously to denote a *jammed heel*.

Shift: (USA) To 'shift the shoes' is an archaic term for a *reset*.

Shin splints: Laying down of new bone on the anterior surface of the cannon due to minor surface fractures caused by stress. **Aka:** Bucked shins (USA). **See:** Periostitis; exostosis.

Shoes: See: Horseshoes.

Shoe pullers (USA): **See:** Pincers.

Sickle-hocked: A conformation fault in which the horse stands with his hind *limbs* bent more than normal at the hock, placing the hooves further forward than ideal. **Aka:** Saber-legged.

Side clips: Quarter *clips*. Particularly those placed near the middle of the quarter, on the sides of the *hoof*.

Sinker: A grave case of *founder* in which *laminitis* has destroyed so many of the *laminae* that the bone column is no longer suspended and begins to sink within the hoof.

Sire: A horse's father.

Solar: The bottom aspect of the horse's hoof.

Sound: Describes a horse who is not *lame* and has no conditions or defects likely to lead to lameness in the future.

Spavin: *From Old French* espavent. Any swelling or abnormal growth in or on the hock. A 'bog spavin' is a soft swelling on the *medial* and/or *dorsal* surface of the hock. A 'blood spavin' is an enlarged vein and a harmless *blemish*. A 'bone spavin' is an *exostosis* on any of the tarsal bones. Large bone spavins are called 'jack spavins'. 'Blind' or 'occult spavins' are exostosises not visible on the exterior of the hock.

Speedy cutting: A gait fault which results in the *interference* of *contralateral limbs*, above the fetlock, at the canter or gallop.

Splint: Ossification of the interosseous *ligament* which attaches a *splint bone* to the cannon bone, causing an *exostosis*. Splints are usually caused by uneven stress on the legs of a young horse or by *trauma*. Lameness may be evident during formation. A splint is considered a *blemish*.

Splint bone: Either of the 2 long, slender bones which run along the back of each cannon bone. The inner splint bone is the second metacarpal (MII) in the fore *limbs* and the second metatarsal in the hinds. The outer splint bone is the fourth metacarpal (MIV) in the fore *limbs* and the fourth metatarsal in the hinds.

TERMINOLOGY AND TECHNICAL LANGUAGE

Square toe: *Horseshoe* shaped so that the toe of the shoe does not follow the curve of the *hoof* but cuts across, usually fit with the toe of the hoof extending out over the shoe. Square toed shoes are usually used on hind hooves when the horse over-reaches. They improve *break-over* in front.

Stabbing: Toe-first landing of a hind foot which causes it to stab into soft turf.

Stallion: An entire male horse, usually 4 years or older.

Stalljack: A miniature *anvil* with a built-on stand. Used by *platers* to shape *plates* and *shoes* without setting down the *hoof*. **Aka:** Fitting stand (UK).

Stay apparatus: The configuration of anatomical structures which allow the horse to rest or sleep whilst standing.

Steel: An alloy of *iron* and carbon. The carbon in steel, usually between 0.2 and 2.0% allows it to be hardened and tempered. Modern steels often contain additional elements for other qualities as well.

Sticker: A light, sharp form of *heel calk* often used on the *lateral* side of hind *race horseshoes*.

Stifled: A stifled horse suffers from recurring, temporary immobilisation of the hock due to the locking of the patella. This condition may be corrected through surgery.

Stud: (1. UK) Shortened name for stud farm. (2. USA) A stallion used for breeding.

Sulcus: Shallow groove on a flat surface, eg central sulcus is the central cleft of the frog.

Superficial crack: Any kind of hoof crack which does not penetrate sensitive tissues. **Aka:** Surface crack.

Superficial digital flexor tendon (SDFT): A *tendon* which runs down the back of the *leg*, splits below the fetlock and attaches to the PI and PII.

Suspensory ligament: A broad band originating at the distal carpal bones, dividing and forming attachment to the proximal sesamoids, before joining the common digital extensor tendon. **Aka:** Suspensory muscle, middle interossous muscle.

Swedge block: (USA) A moulding tool which straps onto the *anvil* and/or fits into its hardie hole. Different swedge blocks can be used to modify *barstock* which can then be forged into rim shoes, polo plates and other shoes with special cross-sections. **Aka:** Swage block.

Swedged shoe: (USA) Any of a number of *horseshoe* styles which have the ground side moulded into a traction modifying pattern. Most feature a deep groove which runs the whole way around the shoe. **Aka:** Full swedged shoe.

Swelled heel (USA): The heel of a *horseshoe* which is folded up onto the hoof surface of the shoe. The hoof surface is then levelled. Swelled heels raise the heels of the hoof without creating as much traction as *blocked heels* or *heel calks*. **Aka:** Wedge heels (UK).

Synovia: See: Synovial fluid.

Synovial fluid: *From Greek* syn *with; and Latin* ovum, *egg*. A very slippery, oil-like substance which is produced by the body to lubricate the joints and *tendons*.

Tarsus: The hock joint.

Tendon: *From Latin* tendo, *to stretch*. Strong fibrous tissue which connects muscle to bone. Tendons function primarily to facilitate movement. Tendons slide within lubricated sheaths, are inelastic and are subject to strains and ruptures.

Tendonitis: Inflammation of a tendon. **See:** Bowed tendon.

Tendosynovitis: Inflammation of the tendon sheath. **See also:** Bowed tendon.

Tenotomy: Surgical severing of a *tendon*.

Therapeutic: Hoof shoeing or trimming done in an attempt to relieve *lameness* or *unsoundness*. **Aka:** Remedial shoeing, pathological shoeing.

Thoracic limb: A fore *limb*.

Thoroughbred: Breed of horse systematically bred for over 300 years for racing.

Thoroughpin: Soft swelling of the *tendon* sheath of the DDFT just above the point of the hock. **Aka:** Through-pin.

Through-pin: See: Thoroughpin.

Thrush: Infection of the tissues of the frog by micro-organisms. This is seen as a foul smelling discharge in the *sulci* and *frog*. Advanced cases may invade sensitive tissues and cause *lameness*.

Toe angle: See: Hoof angle.

Toe extension: Where a shoe or other device continues horizontally beyond the *distal* border of the *hoof wall* in an anterior direction; usually used in foals with *flexural deformities*. **See:** Extension.

Toed-in: The horse's *digit* appears to be twisted inward. This conformation fault usually causes the afflicted *limb* to *paddle*. Horses who are toed in on both fore feet are called pigeon-toed.

Toed-out: The horse's *digit* appears to be twisted outward. This conformation fault usually causes the afflicted *limb* to *dish*. Toed out horses may be prone to *brushing*.

Toe-grab: A form of *toe calk* used on *racing plates*. Toe-grabs are curved with the toe of the shoe and vary in height.

Turf: Grass surface used for *racing*.

Trailer: An extra long heel on a *horseshoe* which is usually turned 45° away from the centre line of the hoof and the line of flight.

Training plate: A very lightweight, usually steel horseshoe used on race horses between races. Training plates are concave fullered (UK) or swedged (USA).

Training shoe: See: Training plate.

Trauma: Injury caused by sudden shock or impact.

Trimming: Removal of excess *horn* using *hoof knife*, *cutters* and *rasp*. **Aka:** Dressing, foot dressing.

True PIII rotation: The *hoof wall* is no longer parallel to the PIII at the toe, PIII is not in alignment with the pastern and the sole is compromised by the tip of the PIII. This occurs in *foundered* horses.

Tungsten: (UK) **See:** Borium.

Turn-down: Where the final ³/₄" (20 mm) of the heel of a *racing plate* is bent distally at approximately 30–40°.

Twitch: Any of several devices used to apply pressure to a horse's upper lip. This is used as form of acupressure of distraction to calm and immobilise the animal.

Two finger radiograph: Reference to the fact that the *dorsal* hoof wall is parallel to the dorsal aspect of the PIII from the coronary band down to an inch or so below, or about the width of 2 fingers. Excluding the extensor process this allows a fairly accurate estimate of the position of the PIII within the hoof, no matter how distorted the lower wall may be.

Under-run heels: The slope of the heels is shallower than that of the toe as viewed from the side. This reduces the *posterior* support of the hoof. Heels need not be short to be under-run. **Aka:** Run-under heels: underslung heels, collapsed heels (UK).

Underslung heels: See: Under-run heels.

Unilateral: On one side only.

Upright foot: *Foot conformation* where the HPA in the mature horse is in excess of 60°; usually *contracted* at the heels; not to be confused with a *club foot*.

Unsoundness: Renders the horse unfit for its intended use.

Valgal: Adjective of *valgus*, eg valgal knees.

Valgus: In to a point, out from that point, eg *carpal valgus*.

Varal: Adjective of *varus*, eg varal knees.

Varus: Out to a point, in from that point, eg *fetlock varus*.

Vein: Vessel which returns blood to the heart.

Wall: The outer, horny part of the *hoof* which grows from the coronary band and is the primary weightbearing structure of the equine *foot*.

Water line: Inner, unpigmented *hoof wall* which is sometimes mistaken for the *white line*.

Web: The width of the stock from which a *horseshoe* is made.

Wedge shoe: A *horseshoe* which is graduated in thickness from a thin toe to thick heels. This has the effect of raising the *hoof angle*. A reverse wedge shoe raises the toe and lowers the *hoof angle*. *Lateral* wedge shoes alter the *mediolateral balance* of the foot. **Aka:** Graduated heel shoe (UK).

Welding, forge: The process of joining metal surfaces by heating them in a *forge* until they are slightly molten, then hammering them together on the *anvil*. Good forge welds are very strong, beadless and may be impossible to detect with the naked eye.

White line: The border between the sole and the *hoof* wall as seen on the *solar* view of the hoof. Usually coloured pale yellow, as opposed to the *water line* which is normally white. Deterioration of the white line and/or inner hoof wall is called white line disease, seedy toe and *onychomycosis*.

Windgall: (UK) **See:** Windpuff.

Windpuff: (USA) A soft swelling which appears on either side of the *fetlock* area. Windpuffs are generally considered blemishes, but may indicate excessive strain which could lead to more serious trouble. 'Articular windpuff' is the distension of the *fetlock* joint capsule. 'Tendinous windpuff' involves the DDFT sheath. **Aka:** Windgall, hygromata; road puff; popped ankle.

Winging-in: See: Dishing.

Winging-out: See: Paddling.

Wry: Distorted *foot*, eg medial heel shunt.

X-rays: X-rays pass through soft tissue and are largely absorbed by bones, they are used in *radiography*. **Aka:** Roentgen rays.

CHAPTER 17

References and further reading

Adams, R. (1990) Non-infectious orthopedic problems. In: *Equine Clinical Neonatatology*. Eds: A.M. Koterba, W.H. Drummond and P.C. Kosch. Lea & Febiger, Philadelphia. pp 333-366.

Butler, K.D. (1985) *The Principles of Horseshoeing II*. Butler Publishing, Maryville.

Curtis, S.J. (1990) The treatment of angular limb deformity by use of medio-lateral extension shoes. *Proc. 2nd Int. Farriery & Lamness Seminar*, Newmarket, England.

Dollar, J.A.W. (1898) *Horseshoeing and the Horse's Foot*. Douglas, Edinburgh.

Hayes, M.H. (1987) *Veterinary Notes for Horse Owners*. 17th edn. Ed: P.D. Rossdale. Stanley Paul, London.

Hickman, J. and Humphrey, M. (Eds) (1988) *Hickman's Farriery*. 2nd edn. J.A. Allen, London.

Leitch, M. (1985) Musculoskeletal disorders in neonatal foals. *Vet. Clin. N. Am.* **1(1)**, 198-199.

Millwater, D. (1995) *The New Dictionary of Farrier Terms and Technical Language*. Dave Millwater Publishing.

Redden, R.F. (1992) A Method of Treating Club Feet – *Proc. 3rd Int. Farriery & Lamness Seminar*; Cambridge, England.

Stashak, T. (Ed) (1987) *Adam's Lameness in Horses*. 4th edn. Lea & Febiger, Philadelphia.

West, G. P. (1995) *Black's Veterinary Dictionary*. 18th edn. A & C Black (Publishers) Ltd, London.

Index

Before seeking references from the index, readers should look up the relevant word in Chapter 17, **Terminology and technical language.** *The index refers only to the main text of the book.*

Abscess, 106, 107, 123, 147, 149, 150, 151, 153
Acrylic, 17, 89, 115, 124

Balance, anterioposterior, 6, 9
Balance, assessment of, 1–3, 5, 7, 91, 93, 95, 97
Balance, medio-lateral, 3, 7, 10, 38, 69, 119
Ballerina syndrome, 43, 45, 46, 58
Barstock, 155, 156, 157, 159
Blocked heels, 85
Borium, 79
Bow-legged, *see carpal varus*
Break-over, 3, 7, 9, 27, 29, 79, 81, 92–94, 97, 98, 103, 107, 109, 111, 113, 117, 139
Breeding, 13, 91
Bruise, 47, 55
Brushing, 92, 99, 101, 103
Bull-nosed foot, 97, 99

Calk, 85, 95
Calkin, 95
Carpus, 27, 34
Check ligament, 7, 9, 48, 51, 53, 57, 95, 97
Clip, 68, 78, 81, 109
Club foot, 5, 9–10, 13, 17, 43–51, 53–54, 69, 71–72, 74, 95, 97, 102
Collapsed heels, 92
Composite, 26, 28, 29, 36–39, 49, 57, 58, 61–64, 84, 85, 100, 102, 115, 120, 121, 125, 127–129

Conformation, 1, 3, 13, 15, 34, 81, 91, 101, 107
Contracted tendons, 13, 71
Corns, 3, 7, 97, 101, 149, 151, 155
Corrective, shoeing and trimming, 1, 15, 17, 53, 145
Coronary band, 2, 3, 45, 47, 57, 93, 105–107, 109, 111, 112, 121
Crack, bar, 107, 108
Crack, deep, 105, 108, 109, 115
Crack, grass, 105–107, 109, 110, 112, 114, 116
Crack, heel, 107, 111, 113
Crack, quarter, 3, 7, 99, 101, 107, 111, 113, 115, 119, 123, 125, 126, 128, 155
Crack, sand, 105, 107, 109, 111, 106, 110, 112, 123, 157
Crack, solar, 107, 108, 115
Crack, superficial, 105, 107, 109
Crack, toe, 107, 108, 111, 117
Cupping, 87

Deep digital flexor tendon (DDFT), 9, 15, 17, 43, 57, 91, 107
Deformity, acquired limb, 91
Deformity, angular limb (ALD), 2, 13, 23, 25, 27–31, 34, 91, 93
Deformity, back at the knee, 73
Deformity, carpal valgus, 3, 14, 25–28, 31, 69, 91, 99
Deformity, carpal varus, 31, 33
Deformity, fetlock valgus, 30, 31, 33, 113
Deformity, fetlock varus, 14, 24, 27,

29–32, 67, 81, 93, 96, 99
Deformity, flexural, 9, 17, 43, 44, 48, 54, 57
Deformity, offset knee, 3, 14, 27, 29–31, 34, 36, 67, 91, 99
Deformity, rotational, 29–31, 69
Desmotomy, 48, 51, 55, 57
Digital hyperextension, 55, 61
Dirt track, 85, 89
Dishing, 99
Dorsal wall resection, 135, 138
Dropped sole, 133, 135–137
Dynamic balance, 2, 5, 7

Environment, 2, 13, 15, 95, 105
Epiphysitis, 28
Expansion, 9, 52, 95, 109, 111, 135
Extension, caudal, 9, 50, 61–64, 65
Extension, medio-lateral, 10, 11, 17, 26–29, 31, 36–40, 96, 103
Extension, toe, 17, 44, 49–51, 53, 55, 57–59
Eyelining, 3, 4, 21, 24, 27, 99

False quarter, 93, 109, 112, 115, 120
False sole, 133
Fire welding, 157, 159
Flaccid tendon, 43, 50, 55, 61, 62, 65
Flat foot, 10, 97, 133
Floating the heel, 98, 101, 155
Foal foot, 16, 17
Foot sore, 131
Forging, 9, 10, 61, 101, 103, 153, 155–157, 159, 160

INDEX

Frog support, 139
Frog trimming, 69

Grooving, 52
Growth plates (epiphyseal),
 14, 21, 23, 25, 27, 34
Growth rings, 2, 45, 73

Hoof angle, 43, 45
Hoof-pastern axis (HPA),
 1, 5–7, 9, 43–55, 59, 67, 70–73, 81,
 92, 95, 97, 99, 111, 132, 151
Hoof wall repair, 51, 73, 82, 97, 100,
 102, 115, 119, 121 *see also patching*
Hoof wall treatments, 73, 151
Hyperextension, 5, 55, 61
Hypoflexion, 43, 55, 61
Hypertrophy, 133, 159, 164

Imbalance, 1, 6, 7, 10, 17, 25, 39, 47,
 53, 69, 91–103, 105, 107, 109–111,
 113, 115, 125, 126, 151, 155, 159
Infection, 93, 105, 107–109, 115,
 123, 125–127, 147, 149, 151, 153
Interfering, 92, 93, 98, 101

Joint laxity, 23, 25

Knock-knees, *see carpal valgus*
Knuckling at fetlock,
 43, 48, 53, 55, 69

Lameness,
 1, 5, 7, 9, 45, 91, 105, 115, 108
Laminitis, 2, 5, 7, 97, 107, 110, 131,
 133, 135, 136, 139, 151, 157
Lesions of the foot,
 5, 32, 63, 93, 104, 109, 115, 116,
 119, 120, 147, 148, 153, 155, 157
Light steel, 83, 87
Long axis, 3, 4, 21, 24, 29
Long-toe, 47, 77, 93, 97, 107, 111
Low-heel, 97, 107

Nailing,
 10, 18, 40, 77, 79, 81, 97, 105, 110
Nailing, French, 115, 118

Opening the heels, 71
Over-reach, 81, 101, 103, 115

Pads, 84, 85, 139, 149, 153
Paddling, 3, 67, 99
Pairing feet, 71
Partial hoof wall avulsion, 115, 155
Patching,
 111, 115, 118, 119, 123–129
Plaiting, 67
Plating, 77, 83, 85, 87, 89
Preying Mantis conformation, 52
Pricking, 18, 83, 147, 149

Restraints, 143
Road work, 67, 71, 77, 78
Rockered toe, 103
Rolled toe,
 73, 81, 113, 117, 161, 164
Rotation, PIII, 45, 47, 53, 91, 135

Safed, 103
Sales, 67–73, 85, 91, 131
Scalping, 81, 98, 101, 103
Seating out, 92, 135, 137, 159, 164
Seedy toe, 10, 97, 105, 109, 116,
 119, 133, 151
Sequestrum, 149, 150
Set toe, 79
Shank, 142, 143, 145
Shelly hooves, 79, 110
Shoe, bar
 6, 9, 86, 98, 101, 113, 120, 155,
 157, 159–161, 165
Shoe, eggbar, 9, 10, 92, 97, 159, 165
Shoe, glue-on, 17, 18, 32, 37, 40,
 49, 50, 61, 85, 88, 135
Shoe, heartbar, 103, 111, 139, 157,
 164, 165

Shoe, hospital plate, 149, 152
Shoe, racing plate, 77, 82–88, 94
Shoe, swan neck, 49, 57, 59
Shoe, swedged, 83, 85
Shoe, three-quarter, 92, 95, 103
Shoe, toe preventer, 101, 103
Shoe, training plate, 77, 85, 87
Shoe loss, 18, 79–81, 101, 115
Side clips, 81, 87
Sinker, 133, 135
Solar view, 3
Speedy cutting, 101, 103
Splints, 27, 29, 99
Square toe, 94, 97
Stalljack, 86
Static balance, 1
Sticker, *see calkin*
Superficial digital
 flexor tendon (SDFT),
 43, 48, 51, 53, 55
Surgical intervention, 25, 26

Thrush, 147, 148
Toe dragging, 73, 79
Toe-in,
 3, 21, 27, 29, 67–69, 93, 99, 113
Toe-out, 17, 21, 69
Toe-grabs, 85, 87, 88, 95
Toe-up conformation, 50, 55, 65
Trailer, 103
Turn-downs, 85, 88
Turf, 77, 87
Twitch, 143, 145

Under-run heels, 3, 7, 9, 17, 87,
 93, 97, 111, 113, 151
Upright foot, 5, 10, 79, 93, 97

Veterinary,
 15, 18, 138, 141, 143, 145

Wedges, 6, 9, 84, 151
Windswept, 32